PHYSICAL Science

Lab Manual

purposeful design publications

Colorado Springs, Colorado

Printed in the United States of America
28 27 26 8 9 10

Physical Science – Lab Manual
Purposeful Design Science series

ISBN 978-1-58331-548-4, Catalog #20093
eISBN 978-1-58331-612-2, Catalog #e20093

Purposeful Design Publications is the publishing division of the Association of Christian Schools International (ACSI) and is committed to the ministry of Christian school education, to enable Christian educators and schools worldwide to effectively prepare students for life. As the publisher of textbooks, trade books, and other educational resources within ACSI, Purposeful Design Publications strives to produce biblically sound materials that reflect Christian scholarship and stewardship and that address the identified needs of Christian schools around the world.

References to books, computer software, and other ancillary resources in this series are not endorsements by ACSI. These materials were selected to provide teachers with additional resources appropriate to the concepts being taught and to promote student understanding and enjoyment.

Purposeful Design Publications
A Division of ACSI
PO Box 62249 • Colorado Springs, CO 80962
800/367-0798 • www.purposefuldesign.com

Name: ______________________________ Date: ____________

Lab 1.1.2A Breaking Water

QUESTION: Is water a compound or an element?

HYPOTHESIS: ______________________________

____

EXPERIMENT:

You will need:	• 2 test tubes	• two 25 cm lengths of insulated copper wire
• 300–400 mL beaker filled with water	• test tube rack	• electrical tape
• 15 g Epsom salts	• 9 V battery	• masking tape
• stirring rod		

Steps:

1. Add the Epsom salts to the water and stir. Some salts may stay at the bottom of the beaker.
2. Completely fill both test tubes with the Epsom salts solution and set them in the test tube rack.
3. Attach one wire to each terminal on the battery and secure them with tape. Be sure the wires make contact with the terminals and the opposite ends of the wires do not touch each other.
4. Place the unattached ends of the wires in the beaker. Do not let the wires touch.

   What do you see? ______________________________

   ____

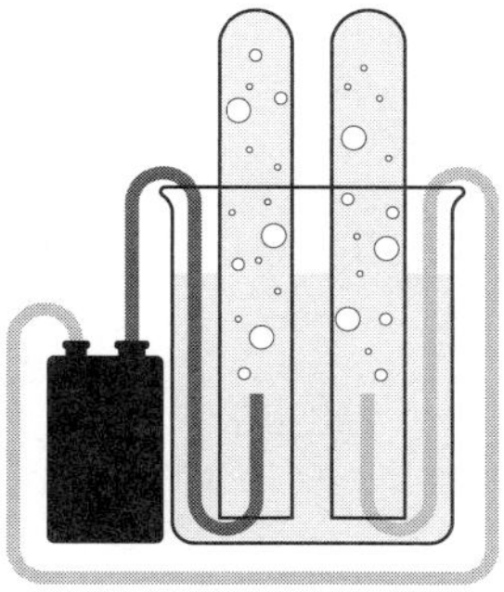

5. Keeping a finger over the opening of one test tube, invert and submerge it, placing the tube over the wire attached to the positive terminal. Repeat with the second test tube, placing it over the wire attached to the negative terminal.
6. Balance the test tubes with the open ends almost touching the bottom of the beaker. Use masking tape to support the test tubes by taping it over the beaker and crossing tape pieces between the two test tubes.
7. Allow the solution to bubble for about 10 minutes.
8. Illustrate the system. Label the positive wire (+), and the negative wire (–). Show the water level in each test tube.

Lab 1.1.2A Breaking Water

ANALYZE AND CONCLUDE:

1. What is the chemical formula for water? ______________________________

2. What two elements make a water molecule? ______________________________

____

3. Which terminal collected the hydrogen gas? Why? ______________________________

____

____

____

4. Which terminal collected the oxygen gas? Why? ______________________________

____

____

____

5. Is water an element or a compound? Why? ______________________________

____

____

____

Name: ________________________________ Date: ______________

Lab 1.1.3A Separating Mixtures

QUESTION: How can sand and salt be separated from water?

HYPOTHESIS: ____

____

EXPERIMENT:

You will need:	• 2 beakers	• hot plate
• sand	• stirring rod	• pan
• salt	• 2 coffee filters	
• paper	• 2 large glass jars	

Steps:

1. Mix sand and salt on a piece of paper. Describe the mixture. ______________

   ____

2. Describe a method for separating the sand from the salt. ______________

   ____

   ____

   ____

3. Add some salt to one beaker and some sand to the other beaker.
4. Add water to both beakers and stir. Illustrate and describe what you see in each beaker. ______________

   ____

   ____

5. Place a coffee filter over the opening of one of the jars. Carefully pour the sand mixture through the filter into the jar. What happened? ______________

   ____

   ____

   ____

Lab 1.1.3A Separating Mixtures

6. Place a clean coffee filter over the opening of the second jar and pour the saltwater through the filter and into the jar.
7. Taste a sample of the water in the jar. How does it taste? ______
8. Describe what happened. ______

ANALYZE AND CONCLUDE:

1. Which mixture was a heterogeneous mixture? ______

2. Which mixture was a homogeneous mixture? ______

3. Which mixture could be separated by a filter? ______

4. Why is the filter able to separate the mixture? ______

5. What process could be used to separate the other mixture? ______

6. Try this process mentioned in your answer and explain what happens. ______

Name: ______________________ Date: __________

Lab 1.1.3B Finding the Iron

QUESTION: How can iron be separated from a mixture?

HYPOTHESIS: ______________________

____

EXPERIMENT:

You will need:	• mass scale	• magnet
• mixture of assorted items		

Steps:

1. Find the mass in grams of the mixture. Record your answer. __________
2. Write a step-by-step procedure for removing all the iron from the mixture.

____

____

____

____

____

____

____

3. Follow the steps in your procedure.
4. Find the mass of the iron and record it here. __________

ANALYZE AND CONCLUDE:

1. What percentage of the original mixture was iron? __________

2. What percentage of the original mixture was not iron? __________

3. Convert the mass of iron from grams to milligrams. __________

4. Convert the mass of the whole mixture from grams to kilograms. __________

____

5. Did your procedure include a step to ensure all the iron was removed from the mixture? If yes, what was it? If no, write that step now. __________

____

____

____

Name: ______________________________ Date: ______________

Lab 1.1.4A Colloids

QUESTION: Which substances are colloids?

HYPOTHESIS: ______________________________

____

EXPERIMENT:

You will need:		
• 5 beakers	• clear soda	• flashlight/laser pointer
• dilute milk solution	• prepared gelatin	• white poster board
	• brewed tea	

Steps:

1. Put each substance in a separate beaker. Add water to the empty beaker.
2. Set up the poster board so the beakers can be placed in front of it.
3. Darken the room.
4. Shine a beam of light through each substance.
5. Record your observations.

Substance	Observations of Light Response
Milk Solution	
Clear Soda	
Prepared Gelatin	
Brewed Tea	
Water	

ANALYZE AND CONCLUDE:

1. Which substances scattered the light? ______________________________

____

2. Which substances did not scatter the light? ______________________________

____

3. Which substances are colloids? Why? ______________________________

____

____

____

Lab 1.1.4A Colloids

4. What is the effect that scatters the light? ______________________

____

5. What type of evidence demonstrates that a substance is a colloid? ______________________

____

____

Name: ______________________________ Date: ______________

Skyscrapers WS 1.1.1A

The chart below lists the 10 tallest skyscrapers in the world. Find the city and country for each building. Then convert the height into meters or feet.

Skyscraper Facts

Name of Skyscraper	City and Country	Height	Conversion
Tapei 101		1,667 ft	
Shanghai Tower		2,073 ft	
Lotte World Tower		554.5 m	
International Commerce Center		484 m	
Makkah Royal Clock Tower		1,972 ft	
Burj Khalifa		828.35 m	
One World Trade Center		541.63 m	
CTF Finance Center		1,739 ft	
Ping An International Finance Center		599 m	
Shanghai World Financial Center		1,614 ft	

1. Which building is the tallest? What is its height in feet? ______________

2. List the skyscrapers in order from shortest to tallest. ______________

____

____

____

3. Which country has most of the tallest skyscrapers? ______________

Name: ________________________________ Date: ______________

Choose Your Science WS 1.1.1B

Use the definitions of chemistry and physics to guide you in deciding which science or sciences would be used in each scenario. Write your answer and explain why.

1. Create a tire that resists skidding and grips the road better during turns. ________

____

2. Design an agent to reduce the flammability of wood. ________________

____

3. Improve a rocket's design so it can be steered more easily. ____________

____

4. Develop a new plastic that can withstand temperatures above 100°C. ________

____

5. Create a device that records colors of specific stars. ________________

____

6. Enhance the adhesive strength and reduce the skin irritability of a bandage. ____

____

7. Study the life cycle of the morel mushroom. ____________________

____

8. Build a pump that carries water from a groundwater source to a mountaintop resort. ________________________________

9. Determine what type of fluid has been found at a crime scene. __________

____

10. Create a patch for a leaking pool that can be applied underwater. ________

____

Research chemistry and physics jobs. List three jobs for each. Then choose one to research in depth. Explain the responsibilities of the job and any current technology that is the result of this type of work.

____

____

____

____

____

Name: ______________________________ Date: ______________

Small Units

WS 1.1.4A

Colloids are heterogeneous mixtures with extremely small particles. Small units define the lengths of small particles. To get an idea of how small micrometers, nanometers, and picometers are, use the relationships given to convert the units.

1 in. = 2.54 cm
1 cm = 10 mm
1 mm = 1,000 micrometers (µm)
1 µm = 1,000 nanometers (nm)
1 nm = 1,000 picometers (pm)

Example: 36 in. = ________ cm $\frac{36 \text{ in.}}{1} \times \frac{2.54 \text{ cm}}{1 \text{ in.}} = 91.44 \text{ cm}$

Example: 36 cm = ________ in. $\frac{36 \text{ cm}}{1} \times \frac{1 \text{ in.}}{2.54 \text{ cm}} = 14.17 \text{ in.}$

1. 655.3 cm = ________ in.

2. 45 mm = ________ µm

3. 83 cm = ________ mm

4. 311 µm = ________ nm

5. 869 nm = ________ pm

6. 7,045 pm = ________ nm = ________ µm

7. 84,814 nm = ________ µm = ________ mm = ________ cm

8. 88.6 pm = ________ µm

9. 99,350 nm = ________ mm

10. 27.3 in. = ________ µm

Measure 10 items that are less than 15 mm in length. Record the name of each item and its length in millimeters. Convert the lengths into micrometers and nanometers.

Small Stuff

Item	Length in mm	Length in µm	Length in nm

Name: ____________________ Date: __________

Lab 1.2.1A Properties

QUESTION: Which property can be used to identify a substance?

HYPOTHESIS: ______________________________

____

EXPERIMENT:

You will need:
- mass scale
- 10 items no larger than your hand and with a mass ≤ 200 g
- metric ruler
- 2 graduated cylinders

Steps:

1. Collect various items.
2. Use the mass scale to measure the mass of each item in grams and record the masses in the data table.
3. For objects with a rectangular solid shape, use a metric ruler to determine the length, width, and height. Use the formula: $V = l \times w \times h$ to determine the volume in cm^3. Record the volumes in the data table.
4. For irregularly shaped objects that will not dissolve in water, determine the volume of the objects by using the water displacement method. Record the volumes in the data table. (1 mL = 1 cm^3)
5. For granulated solids or liquids, use a graduated cylinder to determine the volumes. Record.
6. Use the formula $D = m/V$ to find the density of each object.
7. Divide one of the objects in half. Determine the mass and volume of one of the halves. ______________________________
8. Calculate the densities. ______________________________

Properties

Item	Mass	Volume	Density

Lab 1.2.1A Properties

ANALYZE AND CONCLUDE:

1. Which item had the greatest mass? ______________________________

2. Which item had the greatest volume? ______________________________

3. List the items in order from least dense to most dense. ______________________________

____

____

____

4. How does the density of the divided substance compare with its density before it was divided? Why? ______________________________

____

5. Which property can be used to identify a substance: mass, volume, or density? Why? ______________________________

____

____

Name: ______________________________ Date: ______________

Lab 1.2.1B Comparing Densities

QUESTION: What will happen when two cans of different sodas are placed in water?

HYPOTHESIS: ______________________________

____

EXPERIMENT:

You will need:	• 2 beakers	• 1 can of artificially sweetened soda
• granulated sugar	• mass scale	• large tub
• granulated artificial sweetener	• 1 can of sugar-sweetened soda	

Steps:

1. Add 200 cm^3 of granulated sugar to one beaker.
2. Add 200 cm^3 of granulated artificial sweetener to the second beaker.
3. Predict which substance has a greater mass. ______________________________
4. Measure the mass of each substance and record. ______________________________
5. Fill the large tub with water.
6. Predict what will happen when the can of sugar-sweetened soda is placed in the water. ______________________________
7. Predict what will happen when the can of artificially sweetened soda is placed in the water. ______________________________
8. Place the can of sugar-sweetened soda in the water. Record results. ______________________________
9. Place the can of artificially sweetened soda in the water and record results. ______________________________

ANALYZE AND CONCLUDE:

1. Describe how each sweetener looks. ______________________________

2. Do the descriptions in the previous question discuss characteristic or noncharacteristic properties? Explain the difference. ______________________________

Lab 1.2.1B Comparing Densities

3. Calculate the density of each sweetener. Record here. ______________________

____

____

4. Is density a characteristic or noncharacteristic property? Why? ______________

____

____

____

5. How do your results compare with your predictions? ______________________

____

____

____

____

6. How does the density of the two sweeteners compare with the density of water?

____

____

____

____

Name: ______________________ Date: ____________

Lab 1.2.2A Molecular Motion

QUESTION: How do scientists know that molecules are in motion?

HYPOTHESIS: ______________________

____

EXPERIMENT:

You will need:	• homogenized whole milk	• microscope slide
• beaker	• stirring rod	• coverslip
• distilled water	• pipette	• compound microscope

Steps:

1. In the beaker, stir together a solution of 10 parts distilled water and 1 part milk.
2. Use the pipette to place a drop of the solution on the microscope slide.
3. Carefully set a coverslip over the solution. Use a pencil eraser to gently tap out any air bubbles.
4. Place the slide on the microscope stage and view the sample on lowest power.
5. Record and illustrate your observations. Discuss the shape, size, and motion of what you observe. ______________________

   ____

   ____

   ____
6. Wait for the current to slow down, and then change to medium magnification.
7. Record observations. ______________________

   ____

   ____
8. Before changing to the highest power objective, predict what you will see.

   ____
9. Record observations. ______________________

   ____

Lab 1.2.2A Molecular Motion

ANALYZE AND CONCLUDE:

1. Did you see any movement on the lowest objective? ______________________

2. Why? ______________________

____

3. What substances are in the solution? ______________________

____

____

4. How do the observations on the lowest objective compare with those on the highest objective? ______________________

____

____

____

5. What are the three assumptions of the kinetic theory? ______________________

____

____

____

____

____

6. Do your observations support the kinetic theory? Why? ______________________

____

____

____

____

____

Name: ________________________ Date: __________

Lab 1.2.2B Thermal Expansion

QUESTION: How is air affected by thermal expansion?

HYPOTHESIS: ______________________________

____

Experiment:

You will need:	• ice	• 9" balloon
• large tub	• 1 L plastic bottle	

Steps:

1. Fill the tub about 3/4 full with water and ice.
2. Remove the lid from the bottle and hold the bottle in the water for about 5 minutes. Submerge as much of the bottle as possible without letting any water get inside the bottle.
3. Stand the bottle upright in the tub.
4. Place the balloon over the opening of the bottle, and be sure the seal is airtight.
5. Illustrate the bottle and the balloon. Write a description. ______________________________

   ____

   ____
6. Set the bottle with the balloon aside. Empty the tub.
7. Fill the tub 3/4 full with hot water.
8. Stand the bottle in the tub for 5 minutes.
9. Record and illustrate your observations. ______________________________

   ____

   ____

Bottle and Balloon in Cold Water	Bottle and Balloon in Hot Water

Lab 1.2.2B Thermal Expansion

ANALYZE AND CONCLUDE:

1. What is inside the bottle and the balloon? ______

2. What did the balloon look like after the bottle spent 5 minutes in the ice water?
Why? ______

3. What did the balloon look like after the bottle spent 5 minutes in the hot water?
Why? ______

4. Do your results support the idea of thermal expansion? Why? ______

5. Are your results an example of direct or indirect evidence? Why? ______

6. Write a different experiment that tests thermal expansion using an inflated balloon, a freezer, and sunlight. ______

Name: ______________________ Date: __________

Lab 1.2.3A Crystalline Solids

QUESTION: How do scientists know that alum and salt are crystalline solids?

HYPOTHESIS: ______________________

____

EXPERIMENT:

You will need:		
• 4 beakers	• 2 stirring rods	• 125 g salt
• hot plate	• marking pen	• metric ruler
• 100 g alum	• 1 m thread	• hand lens
	• 4 metal washers	• stereo microscope

Steps:

1. Heat 500 mL of water in a beaker, and dissolve about 100 g alum in the water. Stir as needed.
2. Pour this solution into a clean beaker to a depth of 2 cm. Label the beaker *Alum*.
3. Measure a length of thread that reaches from the bottom of the beaker, over the lip, and onto the table next to it. Tie a washer on either end of the thread to weigh it down.
4. Place the weighted end of the thread in the bottom of the beaker and the other end on the table next to the beaker.
5. In a third beaker, heat 250 mL of water. Dissolve as much salt as possible in it, about 125 g.
6. Pour this solution into the fourth beaker to a depth of 2 cm. Label the beaker *Salt*.
7. Measure a length of thread that reaches from the bottom of the beaker, over the lip, and onto the table next to it. Tie a washer on either end of the thread.
8. Place the weighted end of the thread in the bottom of the beaker and the other end on the table next to the beaker.
9. When the crystals have formed, examine them with a hand lens. Record the size, shape, and color of both sets of crystals. ______________________

    ____

    ____

    ____

10. Repeat Step 9 using a stereo microscope. Place a metric ruler next to the crystals. Record the size, shape, and color of both sets of crystals. ______________________

    ____

    ____

    ____

Lab 1.2.3A Crystalline Solids

11. Illustrate the crystals as seen with both a hand lens and the stereo microscope.

ANALYZE AND CONCLUDE:

1. What is a crystalline solid? ______

2. Did you observe crystalline solids? How do you know? ______

3. Compare the alum and the salt crystals. How are they the same? How are they different? ______

4. What added details could you see through the stereo microscope that you could not see with the hand lens? ______

5. How are crystalline solids different from amorphous solids? ______

6. Why is the structure of a crystalline solid a characteristic property? ______

Name: ______________________________ Date: ______________

Lab 1.2.4A Viscosity

QUESTION: Which liquid has the greatest viscosity?

HYPOTHESIS: ______________________________

____

EXPERIMENT:

You will need:
- 5 graduated cylinders
- mass scale
- transparent shampoo
- maple syrup
- honey
- corn syrup
- metric ruler
- 5 marbles
- 1 large marble
- timer

Steps:

1. If you have a digital mass scale with a tare function, use it to calibrate for the mass of the empty graduated cylinder.
2. If you do not have a digital mass scale, find the density ($D = m/V$) of the empty graduated cylinder and record here. ______________________________
3. Fill each graduated cylinder with 100 mL of a liquid. Find the mass and calculate the density of all five liquids. If you completed Step 2, subtract the density of the cylinder from your answers and record the results in the data table.
4. Measure the height of the liquid and record.
5. Practice starting, stopping, and resetting the timer.
6. Drop a marble into one of the liquids and time how long it takes for the marble to reach the bottom. Record the amount of time.
7. Repeat Step 6 for the other liquids.

Viscosity

Liquid	Density ($D = m/V$)	Height	Time

Lab 1.2.4A Viscosity

ANALYZE AND CONCLUDE:

1. What is the relationship between the viscosity of the liquid and the speed of the marble? ______________________________

____

2. List all the liquids in order, from highest to lowest viscosity. ______________________________

____

____

____

3. How do your results compare with your hypothesis? ______________________________

____

____

4. Do you think a larger marble will travel through the liquids faster than a smaller marble? Why? ______________________________

____

____

5. Test your hypothesis. Choose a liquid to test and remove the marble from the graduated cylinder. You may have to replace some of the liquid that was lost during the removal. How do your results compare with your hypothesis?

____

____

____

6. Why do you think the results are different with a different-sized marble?

____

____

____

Name: ______________________________ Date: ______________

Lab 1.2.4B Viscosity and Temperature

QUESTION: How will a change in temperature affect the viscosity of a liquid?

HYPOTHESIS: ______________________________

____

EXPERIMENT:

You will need:	• 2 thermometers	• 4 marbles
• 4 equal-sized graduated cylinders	• beaker	• timer
• corn syrup	• hot plate	

Steps:

1. Pour 100 mL of corn syrup in a graduated cylinder and place it in the refrigerator.
2. Pour 100 mL of corn syrup in a second graduated cylinder and record its temperature. ______________________________
3. Drop a marble in the second graduated cylinder and measure the time it takes for it to reach the bottom. ______________________________
4. Heat 200 mL of corn syrup to 20° above room temperature. Record the temperature here. ______________________________
5. When the corn syrup has reached the desired temperature, carefully pour 100 mL of corn syrup into a third graduated cylinder. Measure the temperature again and record it. ______________________________
6. Drop a marble in the corn syrup and record the time it takes to reach the bottom. ______________________________
7. Continue heating the remainder of the corn syrup to 40° above room temperature.
8. When the corn syrup has reached the desired temperature, carefully pour 100 mL of corn syrup into the fourth graduated cylinder. Measure the temperature again and record it. ______________________________
9. Drop a marble in the corn syrup and record the time it takes to reach the bottom. ______________________________
10. Take the first graduated cylinder out of the refrigerator. Measure the temperature and record it. ______________________________
11. Drop a marble in the corn syrup and record the time it takes to reach the bottom. ______________________________

Lab 1.2.4B Viscosity and Temperature

12. Using the data you have collected, create a line graph showing the relationship of temperature to the time it takes for the marble to travel through the corn syrup.

ANALYZE AND CONCLUDE:

1. What is the independent variable? On which axis is it placed?

2. What is the dependent variable? On which axis is it placed?

3. What is viscosity?

4. Does change in temperature affect the viscosity of corn syrup?

5. Make a generalization about the viscosity of corn syrup as the temperature increases.

6. Do you think the same generalization can be made for all liquids? Why?

7. Write a lab to test your theory.

Name: ______________________ Date: ____________

Lab 1.2.4C Space Between Molecules

QUESTION: How do scientists know there is space between molecules?

HYPOTHESIS: ____

____

EXPERIMENT:

You will need:
- plastic cup
- tray
- teaspoon
- salt
- sugar
- baking soda

Steps:
1. Set the cup in the tray.
2. Fill the cup with water just to overflowing. When you can see the surface of the water just above the rim of the cup, it is full.
3. Carefully pour 1 tsp of salt in the water.
4. Pour another teaspoon of salt in the water.
5. Continue pouring 1 teaspoon of salt at a time into the water.
6. When you see the water overflow, stop.
7. Record the number of teaspoons you were able to pour into the water.
8. Rinse out the cup.
9. Repeat Steps 2–7 with the sugar and then the baking soda.

Substance	# of Teaspoons

ANALYZE AND CONCLUDE:

1. Is there space between the molecules of water? ____________

2. How do you know? ____

____

____

3. Is the evidence for space between molecules direct or indirect? Why? ____________

____

____

4. Which substance had the greatest number of teaspoons poured into the water?

Why? ____

____

Lab 1.2.4C Space Between Molecules

5. Which substance had the least number of teaspoons poured into the water? Why?

____

____

____

6. Do you think using another liquid would change the results of the lab? Why? ______

____

____

7. Perform the lab again with a different liquid and test your hypothesis. What liquid did you use? ______________________________

Substance	# of Teaspoons

8. How do your results compare to the original lab? Why? ______________

____

____

____

____

____

Name: ______________________________ Date: ______________

Lab 1.2.5A Density

QUESTION: Which gas is the most dense?

HYPOTHESIS: ____

____

EXPERIMENT:

You will need:		
• two 1 L bottles	• 1 tsp yeast	• baking soda
• 450 mL hydrogen peroxide	• 3 balloons	• mass scale
	• vinegar	• permanent marker

Steps:

1. Fill a 1-liter bottle with about 450 mL of hydrogen peroxide.
2. Quickly add about a teaspoon of yeast to the hydrogen peroxide.
3. Immediately cover the opening of the bottle with a balloon. Be sure the seal is airtight.
4. Gently shake the bottle. Bubbles of oxygen (O_2) should begin forming.
5. Let the bottle sit for a few minutes.
6. Take the second bottle and fill it about 1/3 of the way with vinegar.
7. Measure one teaspoon of baking soda and add it to the vinegar. Let this sit for a couple of minutes. The reaction is producing carbon dioxide (CO_2), which is pushing the other air out of the bottle.
8. Add one more teaspoon of baking soda to the vinegar, being careful not to get any on the lip of the bottle.
9. Quickly stretch the balloon over the opening of the bottle, and let this sit for a couple of minutes.
10. When the balloon collecting oxygen and the balloon collecting carbon dioxide have reached their maximum size, carefully remove the balloons and tie them.
11. Record the mass of each balloon.

Density of Gases

Gas	Mass	Volume ($V = 4/3 \times \pi \times r^3$)	Density ($D = m/V$)

12. Take the third balloon and inflate it so it is about the size of the other two balloons. Record the mass.
13. Label each balloon according to the gas it contains: O_2, CO_2, and *Air*.
14. Estimate the volume of each balloon. ($V = 4/3 \times \pi \times r^3$)
15. Calculate the density. ($D = m/V$)

Lab 1.2.5A Density

16. If you were to drop all three balloons at the same time from the height of the ceiling, hypothesize the order in which each balloon would hit the ground.

17. Test your hypothesis. Write the order the balloons landed from first to last.

ANALYZE AND CONCLUDE:

1. What does density measure?

2. According to your calculations, which gas is the most dense?

3. According to your experiment, which gas is the most dense?

4. If your answers for Questions 2 and 3 are not the same, explain what errors may have occurred.

5. What changes could be made to the experiment to reduce experimental errors?

6. If the gases in the experiment were different temperatures, how would that change the results?

Name: ______________________________ Date: ______________

Lab 1.2.6A Boiling Points

QUESTION: Which liquid has the highest boiling point?

HYPOTHESIS: ______________________________

____

EXPERIMENT:

You will need:	• saltwater	• hot plate
• distilled water	• syrup	• heat-resistant gloves
• fruit juice	• 4 beakers	• 4 thermometers

Steps:

1. Pour an equal amount of each liquid into the four beakers.
2. Record the starting temperature of each liquid.
3. Hypothesize the relative boiling points of the liquids. List them in order from lowest boiling point to highest boiling point. ______________________________

   ____
4. Heat each liquid to boiling and record the temperature when the liquid is boiling and maintaining a constant temperature.

Liquid	Starting Temperature (°C)	Boiling Point (°C)

ANALYZE AND CONCLUDE:

1. List the liquids in order from lowest to highest boiling point. ______________________________

____

____

2. How do your results compare with your hypothesis? ______________________________

____

____

Lab 1.2.6A Boiling Points

3. Define *boiling point.* ______________________________

____

____

____

4. How do the boiling points compare with the boiling point of water? Why? __________

____

____

5. Compare the boiling points of the liquids to boiling points of solids. __________

____

____

____

____

6. How do you think the boiling points would change if the liquids were tested at an altitude of 1,850 m? Why? ______________________________

____

____

____

____

7. What type of property would you conclude that a melting point is—characteristic or noncharacteristic? ______________________________

____

____

Name: ______________________________ Date: ______________

Lab 1.2.6B Freezing Points

QUESTION: Which liquid has the lowest freezing point?

HYPOTHESIS: ______________________________

____

EXPERIMENT:

You will need:
- 3 test tubes
- marking pen
- distilled water
- saltwater
- 70% isopropyl alcohol
- 3 rubber stoppers
- 3 temperature probes
- heavy gloves
- tongs
- 3 beakers
- dry ice
- timers

Steps:
1. Label the test tubes *Distilled Water*, *Saltwater*, and *Alcohol*.
2. Fill each test tube with one of the liquids.
3. Place a temperature probe through each rubber stopper and cap the test tubes. Be sure the end of the probe is in the liquid and not touching the side of the test .
4. Record the temperature of each liquid in degrees Celsius.
5. Put on heavy gloves and use tongs to fill three beakers with dry ice. DO NOT TOUCH THE DRY ICE WITH BARE SKIN.
6. Set each test tube in a beaker of dry ice.
7. Record the temperature every 30 seconds until the temperature remains constant. Check sample periodically to see if it is frozen.

Time	Distilled Water	Saltwater	Alcohol
0:00			
0:30			
1:00			
1:30			
2:00			
2:30			
3:00			
3:30			
4:00			
4:30			
5:00			
5:30			
6:00			
6:30			
7:00			

Time	Distilled Water	Saltwater	Alcohol
7:30			
8:00			
8:30			
9:00			
9:30			
10:00			
10:30			
11:00			
11:30			
12:00			
12:30			
13:00			
13:30			
14:00			
14:30			

Lab 1.2.6B Freezing Points

8. Create a line graph showing temperature changes for each liquid.

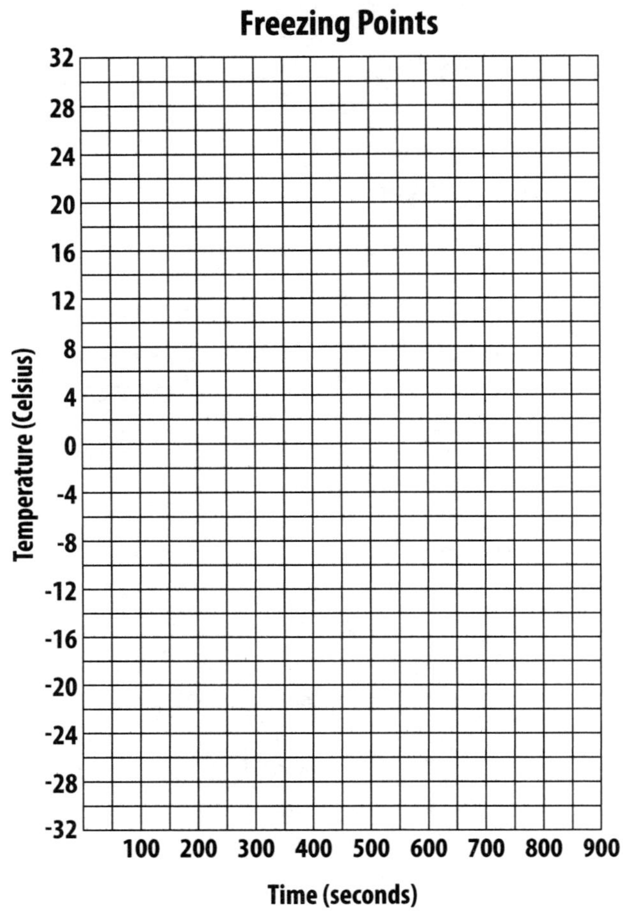

ANALYZE AND CONCLUDE:

1. Define *freezing point.* ______________________

____

____

2. What happened to the alcohol? Why?

____

____

____

____

3. What are the freezing points of distilled water and saltwater? ______________________

____

____

____

____

4. How do your results compare with your hypothesis? ______________________

____

____

5. Why did the liquids freeze? ______________________

____

____

____

6. Which liquid changed temperature the fastest? ______________________

7. Which liquid took the longest time to freeze? Why? ______________________

____

____

8. Is freezing point a characteristic or noncharacteristic property? ______________________

Name: ______________________________ Date: ______________

Lab 1.2.7A Comparing Chemical Reactions

QUESTION: Why will the two iron chlorides react differently with the ammonia?

HYPOTHESIS: ______________________________

____

EXPERIMENT:

You will need:	• 100 mL iron(III) chloride solution	• 10 mL ammonia water
• safety goggles	• 2 Erlenmeyer flasks	
• 100 mL iron(II) chloride solution	• pipette	

Steps:

1. Put on the safety goggles.
2. Pour 100 mL of iron(II) chloride into an Erlenmeyer flask.
3. Add a few drops of ammonia.
4. Swirl the flask briefly and record the results.

   ____

   ____

5. Pour 100 mL of iron(III) chloride into the second Erlenmeyer flask.
6. Swirl the flask briefly and record the results. ______________________________

   ____

   ____

ANALYZE AND CONCLUDE:

1. Did the two types of iron chloride react similarly with the ammonia? ______________

2. What did you observe in the first flask? ______________________________

____

3. What did you observe in the second flask? ______________________________

____

4. Were these chemical changes? Why? ______________________________

____

5. What is a chemical change? ______________________________

____

____

Name: ______________________ Date: __________

Calculate Density

WS 1.2.1A

Compute the density in the following problems. Convert units when necessary and write each answer with the unit g/cm^3.

1. There is a substance with a mass of 50 g and it has a volume of 100 cm^3. What is the density of this substance?

2. A cube of silver measures 4 cm long, 4 cm wide, and 4 cm high. What is its volume? It has a mass of 672 g. What is its density?

3. A bar of gold measures 10 cm wide, 1 m long, and 22 mm high. The bar has a mass of 42,460 g. What is the density of the gold?

4. A block of aluminum has a volume of 25 cm^3 and a mass of 67.5 g. What is its density?

5. The mass of 100 mL of ethyl alcohol is 78.9 g. What is its density?

6. Exactly 59.62 mL of mercury is poured into a cylinder. The mercury has a mass of 0.8050 kg. Calculate the density.

7. An 8 g pebble is added to a graduated cylinder containing 30 mL of water. The water rises to the 32.5 mL mark. What is the density of the pebble?

8. A 1.695 kg block of lead is 10 cm long, 5 cm high, and 30 mm wide. Calculate the volume and density of the lead.

Name: ______________________________ Date: ______________

Solids

WS 1.2.3A

Study crystalline and amorphous solids.

1. Observe three crystalline solids under a stereo microscope or hand lens. Illustrate and label your observations.

2. Observe three amorphous solids under a stereo microscope or hand lens. Illustrate and label your observations.

3. Which crystalline structure was the most interesting? Why? ______________

4. Which amorphous structure was the most interesting? Why? ______________

5. Explain the difference between a crystalline structure and an amorphous structure. ______________

Name: ______________________________ Date: ______________

Gas Laws

WS 1.2.5A

Answer the questions. Show your work and include labels when necessary.

1. Write the equation for Boyle's law. Explain Boyle's law in your own words.

____

____

____

2. Write the equation for Charles's law. Explain Charles's law in your own words.

____

____

____

3. If the initial pressure of a gas is 100 kPa, initial volume is 1.0 L, and final pressure is 25 kPa, what is the final volume?

4. If you have 2.0 L of hydrogen gas under 50 kPa of pressure and you increase the pressure to 250 kPa, what will happen to the volume of the gas? What is the new volume?

5. A tank of oxygen has 125 kPa of pressure. The pressure is increased to 200 kPa and the final volume is 0.5 L. What was the initial volume?

6. If you have 4 L of a gas at a temperature of 173 K and increase the temperature to 273 K, what will happen to the volume? What will the new volume be?

7. A helium tank is filled with 2.0 L of helium at a temperature of 223 K. If the temperature is increased to 446 K, will the pressure inside the tank increase or decrease? What will the new volume be?

8. The initial temperature of a gas is 73 K. The temperature is raised to 182.5 K, which gives it a final volume of 2.5 L. What was the initial volume?

Gas Laws

9. Draw a line on each graph that indicates the relationship between the two variables given, according to Boyle's law and Charles's law. Title the graphs and label the axes.

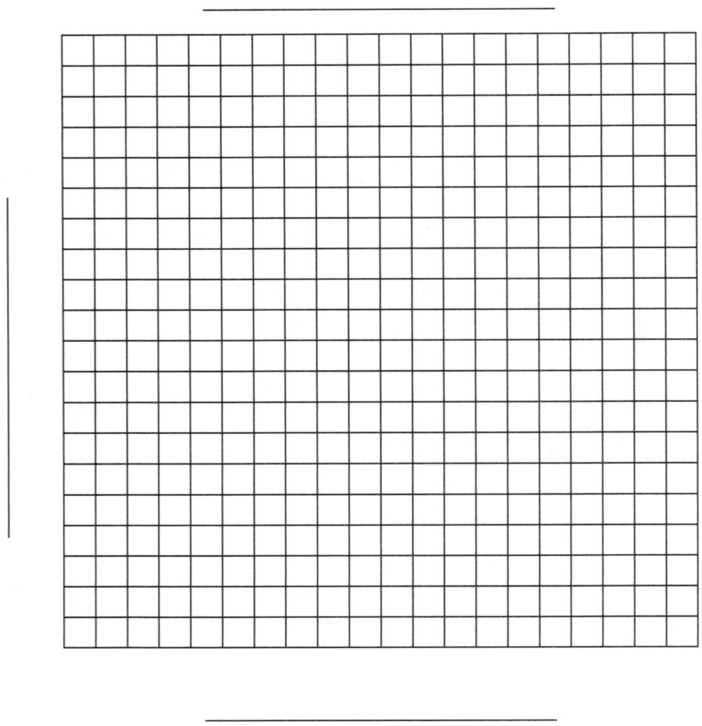

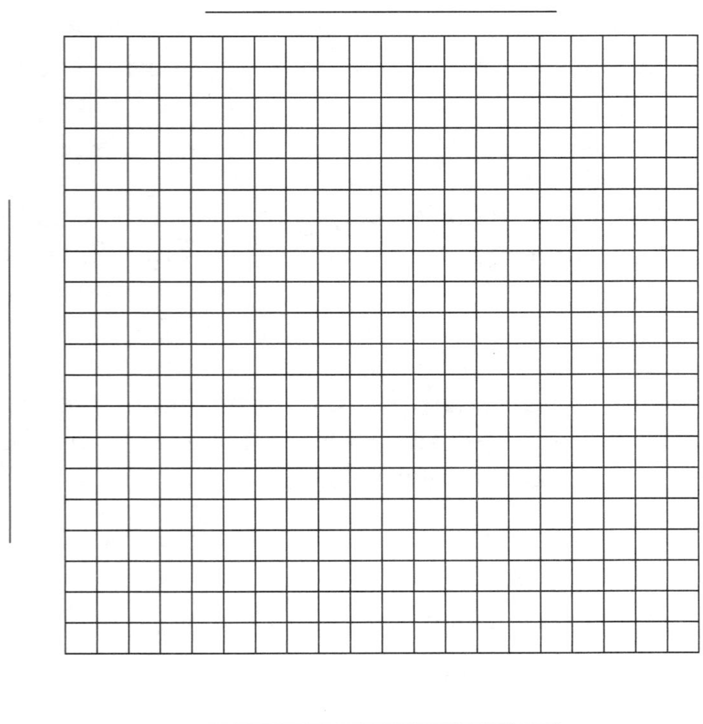

Name: ______________________________ Date: ______________

Atomic Structure

WS 1.3.2A

Fill in the missing components for each neutral atom. Round mass numbers to the nearest whole number.

1.	**Atomic number**		
	Mass number		
	Number of protons	**4**	
	Number of neutrons	**5**	
	Number of electrons	**4**	
	Element name	**beryllium**	
2.	**Atomic number**	**11**	
	Mass number	**23**	
	Number of protons		
	Number of neutrons		
	Number of electrons	**11**	
	Element name	**sodium**	
3.	**Atomic number**		
	Mass number	**70**	
	Number of protons	**31**	
	Number of neutrons		
	Number of electrons		
	Element name	**gallium**	
4.	**Atomic number**		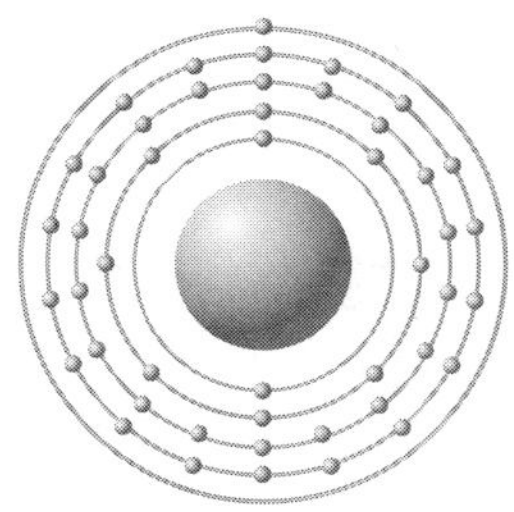
	Mass number	**108**	
	Number of protons	**47**	
	Number of neutrons		
	Number of electrons		
	Element name		
5.	**Atomic number**	**32**	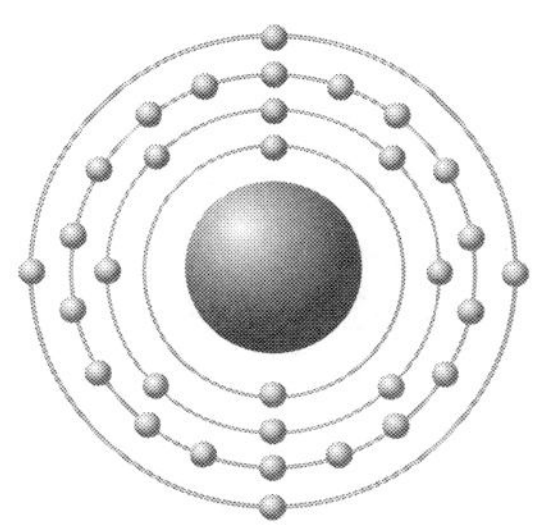
	Mass number		
	Number of protons		
	Number of neutrons		
	Number of electrons		
	Element name		

Name: ______________________________ Date: ______________

Scavenger Hunt WS 1.3.3A

Explore the periodic table using the information you have learned in this lesson. Reference the periodic table in this book to answer each question.

1. How many groups are on the periodic table? ______________________________

2. How many periods are on the periodic table? ______________________________

3. What group and period does titanium belong to? ______________________________

4. What element belongs to group IIIA, period 5? ______________________________

5. Is calcium a metal, nonmetal, or metalloid? ______________________________

6. Is radon a metal, nonmetal, or metalloid? ______________________________

7. Is silicon a metal, nonmetal, or metalloid? ______________________________

8. How many valence electrons does cesium have? Is this element very reactive? Explain. ______________________________

9. How many valence electrons does argon have? Is this element very reactive? Explain. ______________________________

10. How many energy levels does platinum have? ______________________________

Name: ______________________ Date: __________

Lab 1.4.2A Chemical Bonding

QUESTION: How does cobalt(II) chloride react in the presence of other compounds or with the addition of heat?

HYPOTHESIS: ______________________________

____

EXPERIMENT:

You will need:	• spatula	• hot plate
• graduated cylinder	• cobalt(II) chloride hexahydrate	• 250 mL beaker
• test tube	• stirring rod	• tongs
• test tube rack	• 10 mL ethyl alcohol, 95%	

Steps:

1. Pour 10 mL of water into a test tube and place the test tube in the rack. Allow the water to reach room temperature (21°C–23°C). This may take several minutes.
   - Temperature of water: ________
2. Measure approximately 5 g of cobalt(II) chloride hexahydrate and add it to the water in the test tube. Use a stirring rod to gently stir the mixture. Observe the color.
   - Color of water and cobalt(II) chloride mixture: ____________
3. Pour 10 mL of ethyl alcohol into the test tube. Gently mix with the stirring rod. Observe the color.
   - Color of ethyl alcohol and hydrated cobalt(II) chloride mixture: ____________
4. Add an additional 10 mL of water to the test tube. Gently mix with the stirring rod. Observe the color.
   - Color of hydrated cobalt(II) chloride: ____________
5. Prepare a warm water bath by heating a beaker filled with 150 mL of water until it reaches 50°C–55°C. Do not allow the thermometer to touch the bottom or sides of the beaker when measuring the temperature.
6. Place the test tube in the warm water bath for about five minutes.
7. Using tongs, remove the test tube from the water bath and observe the color.
   - Color of the heated anhydrous mixture: ____________

ANALYZE AND CONCLUDE:

1. How does the appearance of anhydrous cobalt(II) chloride (Step 7) and hydrated cobalt(II) chloride (Step 4) compare? ______________________________

____

2. What accounts for the difference in color? ______________________________

Lab 1.4.2A Chemical Bonding

3. How is water added to or removed from the compound? ______

______

______

______

4. What do you think would happen if anhydrous cobalt(II) chloride was exposed to the air over a long period of time? ______

______

______

Name: ______________________ Date: __________

Lab 1.4.3A Bouncing Molecules

QUESTION: How can different bonds affect the physical property of materials?

HYPOTHESIS: ______________________

____

EXPERIMENT:

You will need:	• spatula	• 7.5 g of cornstarch
• hot plate	• 2.5 g of borax	• food coloring
• 100 mL beaker	• stirring rod	• ½ tsp measuring spoon
• 2 disposable foam cups	• graduated cylinder	• tongue depressor
• weighing dish	• 15 mL of white liquid glue	• gloves

Steps:

1. Use the hot plate to heat 35 mL of water in a beaker.
2. Pour the heated water into the first foam cup.
3. Use a weighing dish and spatula to add 2.5 g of borax to the first foam cup.
4. Mix with a stirring rod until well dissolved.
5. Use the graduated cylinder to measure 15 mL of glue and place it in the second foam cup.
6. Use the weighing dish and spatula to add 7.5 g of cornstarch and a few drops of food coloring to the second foam cup. Mix well with a stirring rod.
7. Transfer ½ tsp of the borax and water mixture from the first foam cup into the second foam cup. Wait 15 seconds, and then mix well with a tongue depressor.
8. Squeeze out any excess liquid using glove-covered hands, and roll the mixture into a ball.
9. Drop the ball onto the floor and observe what happens.

ANALYZE AND CONCLUDE:

1. What did the ball do when it was dropped on the floor? ______________________

____

____

2. What types of bonds were formed between the water, borax, glue, and cornstarch?

____

____

3. How do electrons interact in covalent bonds? ______________________

____

____

Name: ______________________________ Date: ______________

Lab 1.4.4A Electricity Conduction

QUESTION: What type of compounds conduct electricity?

HYPOTHESIS: ______________________________

____

EXPERIMENT:

You will need:	• 6 V light bulb and socket	• sugar
• scissors	• three 250 mL beakers	• graduated cylinder
• three 20 cm lengths of insulated wire	• marking pencil	• distilled water
• 6 V dry cell	• sodium chloride	• stirring rod

Steps:

1. Using scissors, carefully remove 2 cm of insulation from each end of the wire pieces.
2. Attach one end of the first wire to one terminal of the dry cell. Connect the other end of the wire to one terminal of the light bulb and socket.
3. Attach one end of the second wire to the other terminal of the dry cell.
4. Attach one end of the third wire to the other terminal of the light bulb and socket.
5. Use the marking pencil to label Beaker 1 *NaCl*, Beaker 2 *sugar*, and Beaker 3 *distilled water.*
6. Place 50 g of sodium chloride in Beaker 1.
7. Add 50 g of sugar to Beaker 2.
8. Holding the insulated portion of each wire, touch the exposed wires to each compound. Record your observations. ______________________________

   ____

   ____

9. Use the graduated cylinder to pour 125 mL of distilled water in Beaker 3.
10. Place the exposed wires in the distilled water and record your observations.

    ____

    ____

11. Add 125 mL of distilled water to Beaker 1 and Beaker 2. Stir to dissolve.
12. Test the conductivity of each solution by placing the exposed wires in the mixtures. Record your observations. ______________________________

    ____

    ____

    ____

Lab 1.4.4A Electricity Conduction

ANALYZE AND CONCLUDE:

1. What compounds or solutions conducted electricity? ______

______

2. What type of compounds conduct electricity? ______

______

3. What type of compounds do not conduct electricity? ______

______

4. Why did some compounds or solutions conduct electricity? ______

______

______

______

______

Name: ______________________________ Date: ______________

Interpreting Chemical Formulas WS 1.4.1A

Determine the identity and quantity of each atom for the compounds in Questions 1–5.

1. SnO_2 ______________________________

2. C_5H_{12} ______________________________

3. $AlCl_3$ ______________________________

4. $CoCO_3$ ______________________________

5. $AgC_2H_3O_2$ ______________________________

Determine the total number of atoms for the compounds in Questions 6–10.

6. H_2SO_4 ______________________________

7. $Ba(NO_3)_2$ ______________________________

8. $(NH_4)_3PO_4$ ______________________________

9. AgCl ______________________________

10. $(NH_4)_2Cr_2O_7$ ______________________________

Write the empirical formulas for the compounds in Questions 11–13.

11. C_2H_2 ______________________________

12. $C_6H_{12}O_6$ ______________________________

13. N_2O_4 ______________________________

Determine the molecular mass of the molecules in Questions 14–15.

14. CCl_4 ______________________________

____

15. HNO_3 ______________________________

____

16. If naphthalene, a compound of carbon and hydrogen, has a molecular mass of 128.174 g and eight hydrogen atoms, how many carbon atoms are in one molecule of naphthalene? ______________________________

____

Name: ______________________________ Date: ______________

Covalent Bonding

WS 1.4.3A

1. Form a bond between two hydrogen atoms. Draw the electron dot diagram below.

2. How many electrons are shared in each bond? ______________________

3. How many total electrons are being shared in the diatomic hydrogen molecule?

____

4. Create a bond between one hydrogen atom and one chlorine atom. Draw the electron dot diagram below.

5. How many total electrons are being shared in the hydrochloric acid molecule?

____

6. Form a bond between one carbon atom and four hydrogen atoms. Draw the electron dot diagram below.

7. How many total electrons are being shared in the carbon tetrahydride molecule?

____

8. Create a bond between one nitrogen atom and three hydrogen atoms. Draw the electron dot diagram below.

9. How many bonds were formed? ______________________

10. How many total electrons are being shared in the nitrogen trihydride molecule?

____

11. How many electrons are not participating in the covalent bond? ______________________

____

12. Draw three other possible molecules.

Name: ______________________ Date: __________

Covalent Molecular Formulas WS 1.4.3B

Name the following covalent molecules:

1. As_2O_3 ______________________

2. CO ______________________

3. CCl_4 ______________________

4. SO_2 ______________________

5. NF_3 ______________________

6. N_2H_4 ______________________

7. ClF_3 ______________________

8. P_2O_3 ______________________

9. S_2F_{10} ______________________

10. N_2O_5 ______________________

Write the formulas for the following covalent molecules.

1. nitrogen monoxide ______________________

2. sulfur trioxide ______________________

3. silicon tetrafluoride ______________________

4. dinitrogen tetrabromide ______________________

5. tetrasulfur tetranitride ______________________

6. phosphorus pentachloride ______________________

7. boron trifluoride ______________________

8. sulfur monoxide ______________________

9. carbon tetrahydride ______________________

10. tricarbon octahydride ______________________

Name: ______________________ Date: __________

Ionic Compound Formulas

WS 1.4.4A

Name the following ionic compounds.

1. NaF ______________________

2. K_2O ______________________

3. Ca_3N_2 ______________________

4. CuO ______________________

5. Cu_2O ______________________

6. Na_2SO_4 ______________________

7. $Be(MnO_4)_2$ ______________________

8. NH_4CN ______________________

9. $Ba(OH)_2 \cdot 2H_2O$ ______________________

10. $LiNO_3 \cdot 4H_2O$ ______________________

Write the formulas for the following ionic compounds.

1. magnesium chloride ______________________

2. rubidium bromide ______________________

3. strontium phosphide ______________________

4. iron(III) nitride ______________________

5. iron(II) nitride ______________________

6. sodium phosphate ______________________

7. calcium carbonate ______________________

8. aluminum carbonate ______________________

9. potassium nitrate pentahydrate ______________________

10. beryllium hypochlorite heptahydrate ______________________

Name: ______________________ Date: __________

Lab 2.1.1A Heat Conduction

QUESTION: Are metals or nonmetals better conductors?

HYPOTHESIS: ______________________

____

EXPERIMENT:

You will need:
- 250 mL beaker
- hot plate
- thermometer
- 20 cm length of wire coat hanger
- 20 cm length of copper wire
- glass stirring rod
- wood dowel
- butter
- spatula

Steps:

1. Fill a 250 mL beaker with approximately 100 mL of water.
2. Place the beaker on the hot plate without turning the hot plate on.
3. In the table below, record the initial temperature of the water. Do not touch the bottom or sides of the beaker with the thermometer.
4. Arrange the piece of coat hanger, copper wire, glass stirring rod, and wood dowel at an angle in the water so the top portion of each object is not exposed to any rising steam when heat is applied. This could affect your results.
5. Using a spatula, apply an equal amount of butter to the end of each object.
6. Slowly heat the water. Ensure the water does not boil over the sides of the beaker.
7. Observe the time and temperature of the water when each butter portion begins to melt. Record these values in the table below.

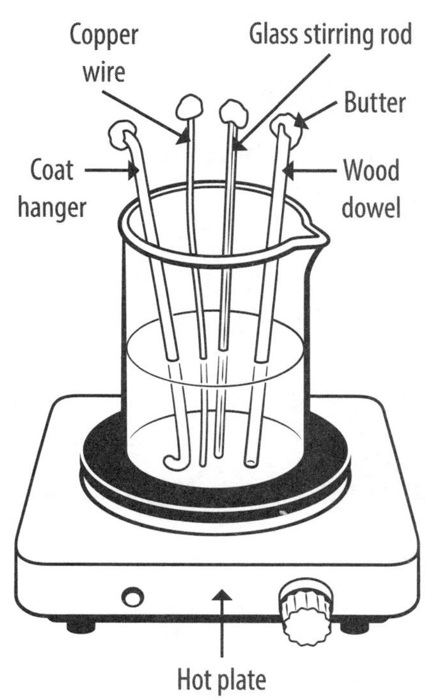

Sample	Time	Temperature (°C)
Initial water temperature	0:00	
Coat hanger		
Copper wire		
Glass stirring rod		
Wood dowel		

Lab 2.1.1A Heat Conduction

ANALYZE AND CONCLUDE:

1. Which heated objects caused the butter to melt first? ______________________

2. Why did the butter samples melt first on these objects? ______________________

3. Which of these materials would make the best thermos material to keep beverages warm? Explain. ______________________

4. Which of these materials would make the best cookware material? Explain. ______

5. Which are better conductors—metals or nonmetals? ______________________

Name: ______________________________ Date: ______________

Lab 2.1.1B Metallic Properties

QUESTION: How are metals and nonmetals different?

HYPOTHESIS: ______________________________

____

EXPERIMENT:

You will need: • watch glass • zinc, carbon, sulfur, magnesium, iron, and copper samples	• scissors • three 20 cm lengths of insulated wire • 6 V dry cell • light bulb and socket	• scale • metric ruler • graduated cylinder

Steps:

1. Place the zinc sample on a watch glass and examine its luster. Describe your observations in the table below.
2. Test the hardness of the zinc sample. Assign a value between 0 and 5 where 0 is not hard and 5 is extremely hard. Note the level of hardness in the table below.
3. Try to bend the zinc sample. If it bends easily, it is malleable. If the sample crumbles and does not bend, it is brittle. Note your observations in the table below.
4. Repeat Steps 1–3 for the carbon, sulfur, magnesium, iron, and copper samples. Document all observations in the table below.

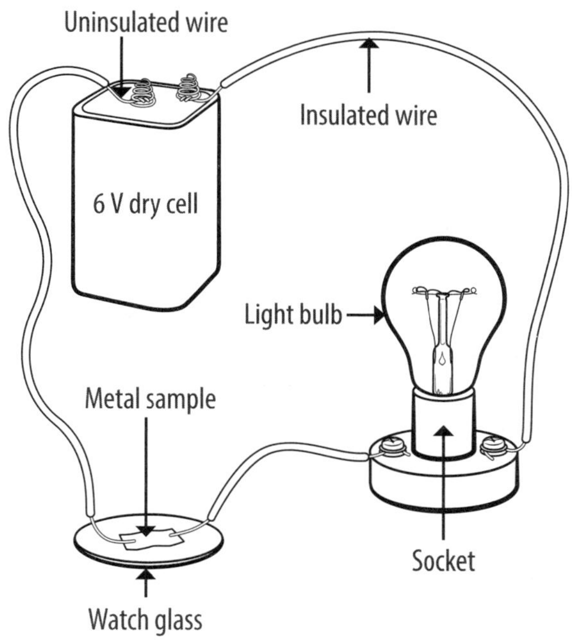

5. Using a pair of scissors, gently remove 1 cm of insulation from each end of the three pieces of wire.
6. Attach one end of wire to a terminal on the dry cell.
7. Connect the other end of the wire to the socket that contains the light bulb.
8. Attach one end of a second wire to the other dry cell terminal.
9. Fasten one end of a third wire to the other portion of the light bulb socket. Ensure that the uninsulated portions of the wires do not come in contact with each other.
10. Holding the insulated portions of the free wires, touch the uninsulated portions of the wires to the zinc sample. If the bulb lights, the material is a good conductor of electricity. Note your observations in the table below.
11. Repeat Step 10 for the carbon, sulfur, magnesium, iron, and copper samples. Record all observations in the table below.

Lab 2.1.1B Metallic Properties

Sample	Luster	Hardness	Malleability	Density (g/mL)	Electrical Conductivity
Zinc					
Carbon					
Sulfur					
Magnesium					
Iron					
Copper					

12. Determine the mass of each sample.
    - Zinc: _____ g
    - Sulfur: _____ g
    - Iron: _____ g
    - Carbon: _____ g
    - Magnesium: _____ g
    - Copper: _____ g
13. For regular-shaped samples, use a metric ruler to determine the volume. The volume of irregular shaped samples can be determined using the water displacement method. Fill a graduated cylinder halfway with water. Note the initial volume. Place the sample in the graduated cylinder and record the volume change.
    - Zinc: _____ mL
    - Sulfur: _____ mL
    - Iron: _____ mL
    - Carbon: _____ mL
    - Magnesium: _____ mL
    - Copper: _____ mL
14. Calculate the density of each sample using the formula $D = m/V$. Record results in the table.

ANALYZE AND CONCLUDE:

1. What properties do metals typically share? ____________________

____

2. What properties do nonmetals typically share? ____________________

____

3. According to your results, what samples are metals? ____________________

____

4. According to your results, what samples are nonmetals? ____________________

____

5. Were there any samples that did not entirely meet the criteria of a metal or a nonmetal? Explain. ____________________

____

6. How do you account for these differences? ____________________

____

Name: ______________________________ Date: ______________

Lab 2.1.2A Single Replacement Reaction

QUESTION: What type of metals create ores?

HYPOTHESIS: ______________________________

____

EXPERIMENT:

You will need:	• dilute HCl solution, 20 mL	• granular zinc
• 250 mL beaker	• graduated cylinder	• tongs
• 10% tin(II) chloride solution, 100 mL	• stirring rod	• sodium bicarbonate

Steps:

1. Pour 100 mL of the tin(II) chloride solution into a beaker.
2. Using a graduated cylinder, measure 20 mL of hydrochloric acid.
3. Slowly add the hydrochloric acid to the tin(II) chloride solution.
4. Gently stir the mixture with a stirring rod.
5. Add about 10 small pieces of granular zinc to the mixture. Note your observations. ______________________________

   ____

   ____
6. Allow the mixture to sit for approximately 5–10 minutes. Describe your observations. ______________________________

   ____

   ____
7. Do not remove the solid until the reaction is complete. Using tongs, gently remove the solid and rinse with cool water. Dispose of the solid in an appropriate trash receptacle. Neutralize the acid solution with sodium bicarbonate and dispose of it according to your teacher's instructions.

ANALYZE AND CONCLUDE:

1. Write the formulas for the two reactions that occurred. ______________________________

____

2. What is the identity of the bubbles that were produced? ______________________________

____

3. Identify the new solid that was formed. ______________________________

4. Why did the new solid float? ______________________________

____

Lab 2.1.2A Single Replacement Reaction

5. According to your observations, is tin or zinc more reactive? Explain. ____________

____

____

____

6. Would a highly reactive metal or less reactive metal readily form an ore in Earth's crust? ____

____

____

Name: ____________________ Date: ____________

Global Ore Distribution

WS 2.1.2A

Research the distribution locations of five main metal ores around the world. Create a key to designate different types of ores, and use the key to note these locations on the map.

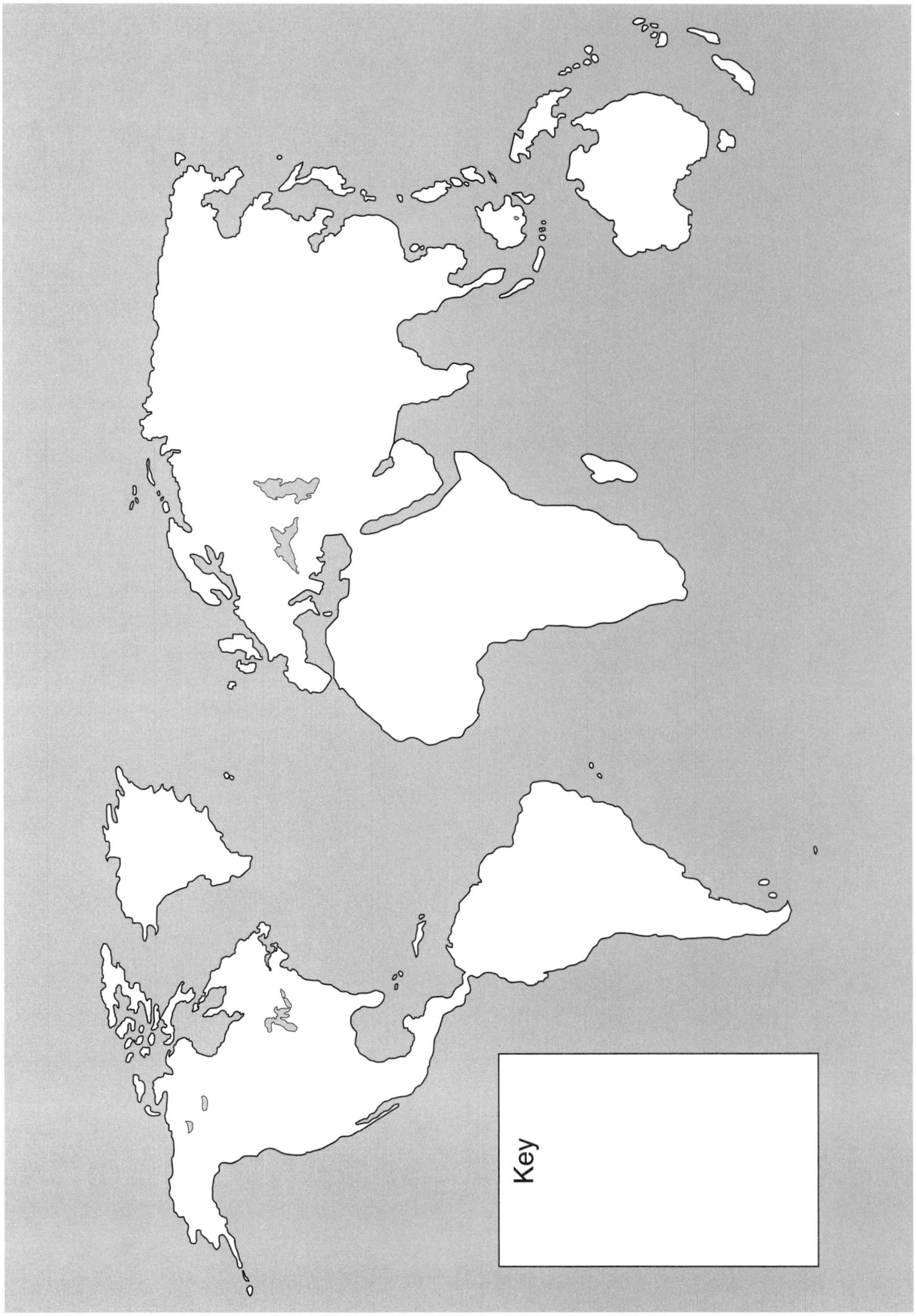

Name: ______________________ Date: ____________

Lab 2.2.2A Ammonia Production

QUESTION: What is hydrogen's main industrial use?

HYPOTHESIS: ______________________

____

EXPERIMENT:

You will need:	• test tube	• 3M sodium hydroxide solution
• 250 mL beaker	• test tube rack	• pipette
• spatula	• graduated cylinder	
• ammonium chloride	• disposable gloves	

Steps:

1. Fill the 250 mL beaker with 150 mL of hot water.
2. With the spatula, add a small scoop of ammonium chloride to the test tube and place it in the rack.
3. Use the graduated cylinder to add about 10 mL of water to the test tube.
4. Wearing disposable gloves, seal the end of the test tube with your thumb and gently mix the solution by inverting the test tube a couple of times.
5. Place the test tube in the beaker of water without spilling any of the contents.
6. Add approximately 10 drops of sodium hydroxide to the test tube with a pipette.
7. Gently swirl the test tube without spilling any of the contents into the beaker.
8. Examine the odor of the product by wafting the scent. Do not directly inhale the product. Note any observations. ______________________

   ____

ANALYZE AND CONCLUDE:

1. What are the formulas for ammonium chloride and sodium hydroxide? ____________

____

2. Predict the formula for ammonia according to the following equation:

$$NH_4Cl_{(aq)} + NaOH_{(aq)} \longrightarrow NaCl_{(aq)} + H_2O_{(l)} + \_\_\_\__{(g)}$$

3. How would you describe the smell of ammonia? ______________________

4. Describe the common industrial process used to produce ammonia. ____________

____

____

5. What are two main industrial uses of ammonia? ______________________

____

Name: ______________________ Date: __________

Lab 2.2.2B Hydrogen Gas

QUESTION: How can you determine if hydrogen gas is present?

HYPOTHESIS: ______________________

____

EXPERIMENT:

You will need:		
• 500 mL beaker	• 2 test tubes	• 2 rubber stoppers
• pipette	• test tube rack	• Bunsen burner
• phenolphthalein solution	• calcium metal	• 2 wood splints
	• funnel	

Steps:

1. Fill the beaker with 400 mL of water.
2. Using a pipette, add 2 drops of phenolphthalein solution to the water.
3. Fill the two test tubes with water and invert the test tubes in the beaker.
4. Add a small piece of calcium metal to the beaker of water.
5. Quickly invert the funnel over the calcium metal. Be sure the funnel is fully submerged.
6. Place one test tube over the neck of the funnel. The opening of the test tube should remain below the surface of the water.
7. When the test tube is filled with gas, remove the test tube and insert a stopper.
8. Place the second test tube over the neck of the funnel and repeat Step 7.
9. Ignite a wood splint with the Bunsen burner and blow out the flame.
10. Hold the glowing splint near the opening of one of the inverted test tubes to check for the presence of hydrogen gas.
11. Repeat with the second test tube if you are unsure about the presence of hydrogen gas.

ANALYZE AND CONCLUDE:

1. What evidence indicated a chemical reaction was occurring? ______________________

____

____

2. Was the initial solution acidic or basic? Was the final solution acidic or basic?

____

3. What did the bubbles represent? ______________________

4. What evidence indicated that hydrogen gas was present in the test tubes?

____

____

Name: ______________________________ Date: ______________

Metal or Nonmetal

WS 2.2.1A

Without referencing the periodic table, determine if the elements are metals or nonmetals according to their property descriptions.

1. This element is found in many diverse regions of the world. It can form more compounds than all the other elements combined. This element is found in coal and petroleum products. It is one of the major components in all plant and animal tissue. This element is a soft solid and a poor conductor of electricity.

   ____

2. This element has one valence electron that it readily donates during a reaction. It is a vital component of most animal and plant life. The appearance of this element is a soft white solid with a silver luster. When burned, it produces a purple flame and a green vapor. It is the seventh most abundant element in Earth's crust. This element reacts vigorously with water.

   ____

3. This silver-gray element is lightweight, extremely strong, and resists corrosion. It is used in many alloys and is found in almost all types of sand and rocks. At high temperatures, this element readily reacts with oxygen and nitrogen. It must be separated from its ore to be useful. This element is very ductile and paramagnetic, and it has a relatively low thermal conductivity.

   ____

4. This element usually exists as a single atom in a gaseous state. It does not normally react with other elements. This element is the most abundant in its family. It is colorless, odorless, and tasteless. Because of its inert nature, this element is useful when welding metals that are highly reactive in oxygen-rich environments.

   ____

5. This highly reactive element has seven valence electrons. It is never found as a free element in nature. This element is essential to your health. Some cultures have found a beneficial use for seaweed that contains a large concentration of this element. It appears as a blue-black solid that readily sublimes to produce a violet vapor.

   ____

6. This element is one of two elements that exists as a liquid at room temperature and standard pressure. It is extremely toxic and in the past was a medical temperature indicator. This element has a silver-white appearance and can freeze into a solid at extremely low temperatures. It is very malleable and conducts heat very well.

   ____

Name: ______________________ Date: __________

Lab 2.3.1A Single and Double Bond Reactions

QUESTION: Are alkanes or alkenes more reactive?

HYPOTHESIS: ______________________

____

EXPERIMENT:

You will need:		
• graduated cylinder	• 2 large test tubes	• pipette
• 40 mL cyclohexane	• test tube rack	• potassium permanganate
	• 40 mL cyclohexene	

Steps:

1. Using a graduated cylinder, measure 40 mL of cyclohexane and pour it into one of the test tubes.
2. Place the test tube in the test tube rack.
3. Using a graduated cylinder, measure 40 mL of cyclohexene and pour it into the second test tube.
4. Place the second test tube in the opposite end of the test tube rack.
5. Using the pipette, add several drops of potassium permanganate to the test tube that contains cyclohexane. Note your observations. ______________________

   ____

6. Using the pipette, add several drops of potassium permanganate to the test tube that contains cyclohexene. Note your observations. ______________________

   ____

ANALYZE AND CONCLUDE:

1. Research the structures of cyclohexane and cyclohexene. How do they compare?

____

____

____

2. What evidence indicates that a reaction took place? ______________________

____

____

3. How does the reactivity of cyclohexane and cyclohexene compare? ______________________

____

____

Name: ____________________ Date: ____________

Isomers

WS 2.3.1A

Determine the isomers for pentane and research the physical properties for each isomer. Record data in the table.

Name	Chemical Formula	Structural Formula	Appearance	Boiling Point	Melting Point
n-pentane					
isopentane					
neopentane					

1. How are the isomers different? ____________________

2. Describe any physical property trends. ____________________

Name: ______________________________ Date: ______________

Lab 2.4.1A Alloy Physical Properties

QUESTION: How do elemental metals and alloys compare?

HYPOTHESIS: ____

____

EXPERIMENT:

You will need:
- samples of brass, bronze, copper, 14-karat gold, pewter, silver, stainless steel, tin, lead, and zinc
- hand lens
- knife or nail
- scale
- metric ruler
- graduated cylinder

Steps:
1. Use a hand lens to examine each sample. Record a description of the appearance of each sample in the table on the back of the page.
2. Using a knife or nail, test the hardness of each sample. Create a hardness scale with values between 0 and 5. Indicate the hardness rating of each sample in the table. If the nail or knife easily scratches the sample the ratio is 0–1.
3. Use a scale to determine the mass of each sample. Record your data in the table.
4. Determine and record the volume of each sample. For regular-shaped objects, measure the length, width, and height of the sample. The volume of irregular shaped objects can be determined using the water displacement method.
5. Calculate and record the density of each sample. Density is equal to mass divided by volume.

ANALYZE AND CONCLUDE:

1. Which sample is the hardest? Which sample is the softest? ______________

____

2. Which sample is the most dense? Which sample is the least dense? ______________

____

3. Identify the samples that are elements. ______________

____

____

4. Identify the samples that are alloys. ______________

____

____

____

____

Lab 2.4.1A Alloy Physical Properties

5. How do elements and alloys compare? ______________________________

____

____

6. Suggest two practical uses for alloys and two practical uses for elements.

____

____

____

____

____

____

Sample	Appearance	Hardness (0–5)	Mass (g)	Volume (cm^3)	Density (g/cm^3)
Brass					
Bronze					
Copper					
14-karat gold					
Pewter					
Silver					
Stainless steel					
Tin					
Lead					
Zinc					

Name: ____ Date: ______________

Lab 2.4.4A Synthesizing Plastic

QUESTION: How can people change the properties of polymers?

HYPOTHESIS: ____

____

EXPERIMENT:

You will need:		
• grater	• weighing dish	• Bunsen burner
• large Russet potato	• spatula	• watch glass
• large mortar and pestle	• balance	• stirring rod
• graduated cylinder	• wax pencil	• pH test strips
• distilled water	• three 250 mL beakers	• 0.1 M sodium hydroxide solution
• mesh strainer	• 0.1 M hydrochloric acid solution	• 2 petri dishes
• 400 mL beaker	• glycerin	
	• ring stand or tripod	

Steps:

1. Grate about 100 g of a clean, large peeled Russet potato.
2. Place the grated potato into a large mortar.
3. Using a graduated cylinder, measure 100 mL of distilled water and pour into the mortar.
4. Gently grind the potato and water mixture with the pestle, creating starch water.
5. Pour the starch water through a strainer into a 400 mL beaker. Do not pour the potato into the strainer.
6. Repeat steps 3 through 5 two times.
7. Allow the starch mixture in the 400 mL beaker to remain undisturbed for at least five minutes.
8. Carefully pour off the water without pouring out the white starch residue that has settled on the bottom of the beaker.
9. Using a graduated cylinder, measure an additional 100 mL of distilled water and pour the water into the 400 mL beaker.
10. Gently stir the water and starch, and allow to remain undisturbed for another five minutes.
11. Pour off the water leaving the starch slurry.
12. Using a weighing dish, a spatula, and the balance, measure 10 g of the starch slurry. With the wax pencil, label one 250 mL beaker *with plastic additive* and label a second 250 mL beaker *without plastic additive*. Place 5 g of the starch slurry into each beaker.
13. To the beaker labeled *with plastic additive*, add 25 mL of water, 5 mL of hydrochloric acid, and 3 mL of glycerin to the beaker.
14. To the beaker labeled *without plastic additive*, add 25 mL of water and 5 mL of hydrochloric acid. Do not add any glycerin.
15. Set up the ring stand and Bunsen burner apparatus according to your teacher's instructions.

Lab 2.4.4A Synthesizing Plastic

16. Place the watch glass over the beaker labeled *with plastic additive* and gently heat the mixture until it begins to boil.
17. Boil the mixture for approximately 15 minutes. If it looks like all of the liquid is about to boil out, remove the Bunsen burner and allow the beaker to cool.
18. Insert the stirring rod into the mixture and then touch the moist end of the rod to a pH test strip.
19. Begin adding small amounts of sodium hydroxide to the starch mixture until the test strips show a neutral status.
20. Repeat steps 16 through 19 for the beaker labeled *without plastic additive*.
21. Transfer an even layer of each starch mixture onto two petri dishes, and allow the mixtures to dry for several days.
22. Observe the two products after several days.

ANALYZE AND CONCLUDE:

1. How do the two products compare? ______________________________

____

____

____

2. What compound determined the physical properties of the two products? Why do you think there are property differences? ______________________________

____

____

3. What type of materials do you think could be made from the product that contains glycerin? ______________________________

____

____

Name: ______________________ Date: ______________

Alloy Components

WS 2.4.1A

Research 10 different alloys and identify the elements that combine to make each alloy. Create a key that includes the name of each alloy and an associated color or symbol of your choosing. Color or insert the symbol in the element boxes on the periodic table that make each alloy. For example, the symbol associated with brass could be a circle. Brass is made of copper and zinc elements. A circle would be added to the copper and zinc element boxes on the periodic table. Remember that many alloys often share the same elemental components.

PERIODIC TABLE OF ELEMENTS

Key:
- ● Brass

	IA 1	IIA 2	IIIB 3	IVB 4	VB 5	VIB 6	VIIB 7	VIIIB 8	9	10	IB 11	IIB 12	IIIA 13	IVA 14	VA 15	VIA 16	VIIA 17	VIIIA 18
1	1 H																	2 He
2	3 Li	4 Be											5 B	6 C	7 N	8 O	9 F	10 Ne
3	11 Na	12 Mg											13 Al	14 Si	15 P	16 S	17 Cl	18 Ar
4	19 K	20 Ca	21 Sc	22 Ti	23 V	24 Cr	25 Mn	26 Fe	27 Co	28 Ni	29 Cu ●	30 Zn ●	31 Ga	32 Ge	33 As	34 Se	35 Br	36 Kr
5	37 Rb	38 Sr	39 Y	40 Zr	41 Nb	42 Mo	43* Tc	44 Ru	45 Rh	46 Pd	47 Ag	48 Cd	49 In	50 Sn	51 Sb	52 Te	53 I	54 Xe
6	55 Cs	56 Ba		72 Hf	73 Ta	74 W	75 Re	76 Os	77 Ir	78 Pt	79 Au	80 Hg	81 Tl	82 Pb	83 Bi	84 Po	85 At	86 Rn
7	87 Fr	88 Ra		104* Rf	105* Db	106* Sg	107* Bh	108* Hs	109* Mt	110* Ds	111* Rg	112* Cn	113* Nh	114* Fl	115* Mc	116* Lv	117* Ts	118* Og

Lanthanides	57 La	58 Ce	59 Pr	60 Nd	61* Pm	62 Sm	63 Eu	64 Gd	65 Tb	66 Dy	67 Ho	68 Er	69 Tm	70 Yb	71 Lu
Actinides	89 Ac	90 Th	91 Pa	92 U	93* Np	94* Pu	95* Am	96* Cm	97* Bk	98* Cf	99* Es	100* Fm	101* Md	102* No	103* Lr

Name: ______________________________ Date: ______________

Household Items

WS 2.4.2A

Conduct a material survey of your home. List objects made of glass or other types of ceramics. Determine what qualities make the materials useful, and determine or research the types of materials that were used to manufacture each object.

Object	Type of Ceramic	Useful Qualities	Types of Materials

Name: ____________________ Date: __________

Recycling Tidbits

WS 2.4.3A

Test your knowledge about recycling by answering the following questions:

1. It takes approximately _____ percent of the energy to recycle an aluminum can as it does to make the original aluminum can.
ⓐ 5 ⓒ 25
ⓑ 10 ⓓ 50

2. The energy saved by recycling an aluminum can will run a computer for _____.
ⓐ one hour ⓒ four hours
ⓑ three hours ⓓ twenty-four hours

3. The energy saved by recycling a glass bottle will run a 100-watt light bulb for _____.
ⓐ thirty minutes ⓒ two hours
ⓑ one hour ⓓ four hours

4. One ton of recycled newsprint saves approximately _____ mature trees.
ⓐ 5 ⓒ 17
ⓑ 10 ⓓ 20

5. It takes approximately _____ recycled 2-liter plastic bottles to make enough polyester fiber to fill one sleeping bag.
ⓐ 35 ⓒ 350
ⓑ 50 ⓓ 500

6. The recycling of all of the newspapers, cardboard, glass, and metal that the average American household throws away every year can reduce carbon dioxide emissions by about _____ per year.
ⓐ 180 kg ⓒ 380 kg
ⓑ 270 kg ⓓ 450 kg

7. Every year, one mature tree filters up to _____ of pollutants from the air.
ⓐ 27 kg ⓒ 99 kg
ⓑ 36 kg ⓓ 120 kg

Name: ______________________________ Date: ______________

Plastic Recycling Codes

WS 2.4.3B

In the United States, a group of individuals developed a numbering system to indicate the type of plastic used in a variety of materials. These numbers can be found on the bottom of most materials that contain plastic. When the plastic material is recycled, the specific recycling method is quickly recognized. Your task is to research the seven resin codes and complete the table below. Provide the full plastic name along with the abbreviation in the second column.

Resin Code	Plastic Name	Traditional Uses	Recycled Products
1 PET			
2 HDPE			
3 V			
4 LDPE			
5 PP			
6 PS			
7 OTHER			

Name: ______________________ Date: __________

Lab 3.1.1A Temperature

QUESTION: How does temperature affect dissolving?

HYPOTHESIS: ______________________

____

EXPERIMENT:

You will need:	• sugar granules	• thermometer
• weighing paper	• marking pencil	• hot plate
• spatula	• three 250 mL beakers	• paper towels
• mass scale	• distilled water	

Steps:

1. Using the weighing paper, spatula, and mass scale, measure three 50 g samples of sugar granules.
2. Using the marking pencil, label the beakers *1*, *2*, and *3*.
3. Pour 100 mL of room-temperature distilled water into each beaker.
4. Record the temperature of the water in Beaker 1.
5. Add sugar from one 50 g sample to Beaker 1 until no more sugar will dissolve and it settles on the bottom. Do not stir the solution.
6. Measure the mass of the remaining sugar, and subtract that number from 50 g to determine the amount of sugar that dissolved. Record your data.
7. On the hot plate, warm the water in Beaker 2 to 50°C.
8. Add sugar from the second sugar sample to Beaker 2 until no more sugar will dissolve and it settles on the bottom. Do not stir the solution.
9. Measure the mass of the remaining sugar, and subtract that number from 50 g to determine the amount of sugar that dissolved. Record your data.
10. Boil the water in Beaker 3, bringing it to a temperature of 100°C.
11. Quickly add sugar from the third sugar sample to the hot water until no more sugar will dissolve and it settles on the bottom. Do not stir the solution.
12. Determine the mass of the remaining sugar, and subtract that number from 50 g to determine the amount of sugar that dissolved. Record your data.

	Temperature (°C)	**Mass of Dissolved Sugar (50 g – mass of remaining sugar)**
Beaker 1		
Beaker 2		
Beaker 3		

Lab 3.1.1A Temperature

ANALYZE AND CONCLUDE:

1. How does temperature affect the amount of sugar dissolved? ______________________

____

____

2. How does the temperature of the solvent affect the dissolving of the solute? ______

____

____

____

____

____

3. Using the lab results, compare the rate of sweetening iced tea to hot tea. _________

____

____

4. How well would sugar dissolve in water that is just above freezing? ____________

____

____

Name: ______________________________ Date: ______________

Lab 3.1.1B Surface Area

QUESTION: How does surface area affect dissolving?

HYPOTHESIS: ______________________________

____

EXPERIMENT:

You will need:	• mass scale	• distilled water
• weighing papers	• mortar and pestle	• stirring rod
• sugar cubes	• marking pencil	• timer
• sugar granules	• three 250 mL beakers	

Steps:

1. Onto weighing papers, measure 50 g of sugar cubes and two different 50 g samples of sugar granules.
2. Using the mortar and pestle, grind one of the 50 g samples of sugar granules into fine powder.
3. Using the marking pencil, label the beakers *1*, *2*, and *3*.
4. Pour 100 mL of distilled water into each 250 mL beaker.
5. Slowly pour the ground sugar into the center of Beaker 1, stirring as you pour. Do this for 30 seconds or until no more sugar will dissolve and any remaining undissolved sugar settles on the bottom. Follow the same procedure with the sugar granules and cubes.
6. Calculate the mass of the remaining ground sugar and sugar granules, and subtract those numbers from 50 g to determine how much of each type of sugar dissolved. If a portion of the sugar cubes remain, pour the liquid through filter paper and retrieve the remaining sugar portion. Allow it to dry before calculating how much dissolved.

	Mass of Dissolved Sugar (50 g – remaining sugar)
Ground Sugar	
Sugar Granules	
Sugar Cubes	

ANALYZE AND CONCLUDE:

1. How do the results compare with the hypothesis? ______________________________

____

2. Which type of sugar dissolved the fastest? ______________________________

3. Which dissolved the slowest? Why? ______________________________

____

____

Lab 3.1.1B Surface Area

4. How would the results be different if there were no time limit? ______________________

____

____

5. How would the results be different if tap water had been used instead of distilled water? ______________________

____

____

____

____

Name: ______________________ Date: __________

Lab 3.1.1C Movement

QUESTION: How does movement affect dissolving?

HYPOTHESIS: ______________________

____

EXPERIMENT:

You will need:
- weighing papers
- sugar granules
- mass scale
- distilled water
- three 250 mL beakers
- marking pencil
- 2 stirring rods

Steps:
1. Onto weighing papers, measure three 50 g samples of sugar granules.
2. Pour 100 mL of distilled water into each 250 mL beaker. With the marking pencil, label the beakers *1*, *2*, and *3*.
3. Slowly add sugar from one of the 50 g samples to Beaker 1 until no more sugar will dissolve and the remaining undissolved sugar settles on the bottom. Do not stir the solution.
4. Add sugar from the second 50 g sample to Beaker 2, stirring slowly until no more sugar will dissolve and the remaining undissolved sugar settles on the bottom.
5. Add sugar from the third 50 g sample to Beaker 3, stirring vigorously until no more sugar will dissolve and the remaining undissolved sugar settles on the bottom.
6. Calculate the mass of the remaining sugar of each sample. Subtract those numbers from 50 g to determine how much of each type of sugar dissolved.
7. Record your data.

	Mass of Dissolved Sugar (50 g – remaining sugar)
Unstirred Solution, Beaker 1	
Slowly Stirred Solution, Beaker 2	
Vigorously Stirred Solution, Beaker 3	

ANALYZE AND CONCLUDE:

1. How do the results compare with the hypothesis? ______________________

____

____

2. In which beaker did the most sugar dissolve? ______________________

____

3. In which beaker did the least sugar dissolve? ______________________

____

Lab 3.1.1C Movement

4. Explain the results. ______________________________

____

____

____

____

5. Producers of packaged juice often advise shaking the container before opening it.

Why? ______________________________

____

____

____

6. Are these juices solutions? Why? ______________________________

____

____

____

Name: ______________________________ Date: ______________

Lab 3.1.1D Polar and Nonpolar

QUESTION: How do solutes interact with polar and nonpolar solvents?

HYPOTHESIS: ______________________________

____

EXPERIMENT:

You will need:
- 12 test tubes
- test tube rack
- spatula
- urea crystals
- iodine crystals
- ammonium chloride
- sodium chloride (table salt)
- 3 pipettes
- ethanol
- vegetable oil
- stirring rod
- paper towels

Steps:

1. Fill 6 of the test tubes three-fourths full of water. Place the test tubes in the test tube rack.
2. Using a spatula, add a few urea crystals to the first test tube, iodine crystals to the second test tube, ammonium chloride crystals to the third test tube, and sodium chloride to the fourth test tube.
3. Using a pipette, add 10 drops of ethanol to the fifth test tube and 10 drops of vegetable oil from a different pipette to the sixth test tube.
4. Gently stir the contents of each test tube. Wipe off the stirring rod with a paper towel after each use.
5. Observe the extent to which the solvent dissolves each solute. Record your observations in the data chart using the key: *S* = soluble, *SS* = slightly soluble, *IN* = insoluble.
6. Using 5 new test tubes and vegetable oil as the solvent, repeat Steps 1–5. Instead of adding vegetable oil to the 12th test tube, add 10 drops of water.

Water as a Solvent

Solute	Solubility (S, SS, or IN)
Urea	
Iodine	
Ammonium Chloride	
Sodium Chloride	
Ethanol	
Vegetable Oil	

Vegetable Oil as a Solvent

Solute	Solubility (S, SS, or IN)
Urea	
Iodine	
Ammonium Chloride	
Sodium Chloride	
Ethanol	
Water	

Lab 3.1.1D Polar and Nonpolar

ANALYZE AND CONCLUDE:

1. Which solutes were more soluble in water than in vegetable oil? ______________________

____

2. Which solutes were more soluble in vegetable oil than in water? ______________________

____

3. Were any substances equally soluble in both water and vegetable oil? Why?

____

____

____

4. Which of the solutes are polar? ____

____

5. Which of the solutes are nonpolar? ____

____

6. Explain the phrase "like dissolves like." ____

____

____

____

Name: ______________________ Date: __________

Lab 3.1.2A Solubility

QUESTION: Which solids are water-soluble?

HYPOTHESIS: ______________________

____

EXPERIMENT:

You will need:
- ten 500 mL beakers
- sugar
- ash
- flour
- powdered drink mix
- cornstarch
- instant coffee
- cocoa powder
- talcum powder
- soap powder
- gelatin
- mortar and pestle
- 10 weighing dishes
- mass scale
- marking pencil
- stirring rod
- paper towels

Steps:

1. Fill each beaker with 300 mL of water.
2. Crush the solids into particles of equal size. After crushing each substance, place it in its own weighing dish.
3. Determine the mass of each substance. Record your data.
4. Label each beaker with a name of a substance.
5. Slowly add each substance to the appropriate beaker, stirring constantly.
6. Clean the stirring rod each time before changing beakers.
7. Determine how much of each substance dissolves in 300 mL of water.
8. Carefully observe the beakers to determine which substances formed solutions and which did not.

Substance	Initial mass	Mass dissolved	Water-soluble?

ANALYZE AND CONCLUDE:

1. What is the definition of a solution? ______________________

____

____

Lab 3.1.2A Solubility

2. Which substances formed aqueous solutions? ______

3. Which substances did not form aqueous solutions? ______

4. Which substances are water-soluble? ______

5. How do the results compare with the hypothesis? ______

6. How do the results compare with other lab groups? ______

7. What errors might have occurred to produce incorrect results? How could the procedure be improved to reduce errors? ______

Name: ______________________ Date: ____________

Lab 3.1.3A Distillation

QUESTION: How can water be separated from salt and food coloring?

HYPOTHESIS: ______________________

____

EXPERIMENT:

You will need:	• hot plate	• 20 cm × 30 cm piece of aluminum foil
• safety goggles	• salt	• bowl
• 1 L beaker	• stirring rod	• salinity test strip
• distilled water	• food coloring	

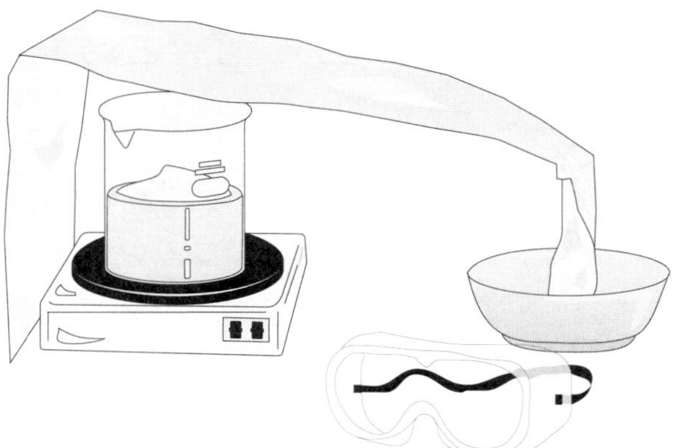

Steps:

1. Put on the safety goggles.
2. Fill the beaker with 500 mL of distilled water and set it on the hot plate to boil.
3. Add salt to the water, stirring with the rod until no more salt will dissolve.
4. Add several drops of food coloring.
5. Fold the aluminum foil into a tent shape with the tall end of the tent 1 cm above the beaker and the short end 4 cm high.
6. Set the aluminum tent over the beaker, and place the bowl beneath its short end.
7. Observe the color of the water that collects in the bowl.
8. Use a salinity test strip to determine the salinity of the water collected in the bowl. Record your data. ______________________

ANALYZE AND CONCLUDE:

1. What type of solution is this? ______________________

2. List the solvent and the solutes. ______________________

____

3. Explain the process to extract pure water from the solution. ______________________

____

____

____

____

Lab 3.1.3A Distillation

4. What is the purpose of the aluminum foil? ______________________________

____

____

____

____

5. What are the challenges in this lab? ______________________________

____

____

____

____

____

6. How can the procedure be changed to overcome those challenges? ______________

____

____

____

____

____

Name: ______________________________ Date: ______________

Lab 3.1.4A Temperature, Solubility, and Concentration

QUESTION: How will different temperatures affect the solubility of sodium thiosulfate?

HYPOTHESIS: ______________________________

____

EXPERIMENT:

You will need:	• sodium thiosulfate ($Na_2S_2O_3$)	• heat-resistant gloves
• safety goggles	• three 250 mL beakers	• stirring rod
• mass scale	• thermometer	• hot plate

Steps:

1. Put on the safety goggles.
2. Determine the mass of the sodium thiosulfate. Record the mass.
3. Fill one beaker with 100 mL of room-temperature water. Measure the temperature and record your data.
4. Wearing heat-resistant gloves, slowly add sodium thiosulfate until no more will dissolve and the crystals begin swirling at the bottom of the beaker. Stir the liquid constantly with the stirring rod as you add the solute.
5. Calculate and record the amount of sodium thiosulfate the room-temperature water dissolved.
6. Pour 100 mL of room-temperature water into the second beaker, and add the same amount of sodium thiosulfate that dissolved in the first beaker.
7. Slowly heat the second beaker on the hot plate, but do not bring the solution to a boil. Heat the solution until small bubbles form, then remove it from the hot plate and allow it to cool slightly.
8. Measure the temperature and record your data.
9. Add more sodium thiosulfate to the near-boiling water, stirring constantly until no more will dissolve.
10. Calculate and record the amount of sodium thiosulfate the near-boiling water dissolved.
11. Pour 100 mL of room-temperature water into the third beaker, and add the same amount of sodium thiosulfate that was dissolved in the first beaker.
12. Bring the solution to a full boil. While it boils, add sodium thiosulfate until no more will dissolve.
13. Calculate and record the amount of sodium thiosulfate the boiling water dissolved.
14. Remove the beaker from the hot plate, and allow the solution to cool.
15. When the second and third solutions have cooled to room temperature, add a small crystal of sodium thiosulfate to each of the three beakers. Record an observation for all three beakers.

Lab 3.1.4A Temperature, Solubility, and Concentration

	Original Mass	Dissolved	Temperature
Beaker 1			
Beaker 2			
Beaker 3			

ANALYZE AND CONCLUDE:

1. What type of solutions were made in each of these beakers? ______________________

____

____

2. What happened to the added crystals from Step 15 in each of the three solutions?

Why? ____

____

____

____

____

3. Graph the solubility of sodium thiosulfate at all three temperatures. What is the independent variable? What is the dependent variable? ______________________

____

____

____

____

4. Is there enough data to determine the solubility at 50°C? Why? ______________________

____

____

____

____

____

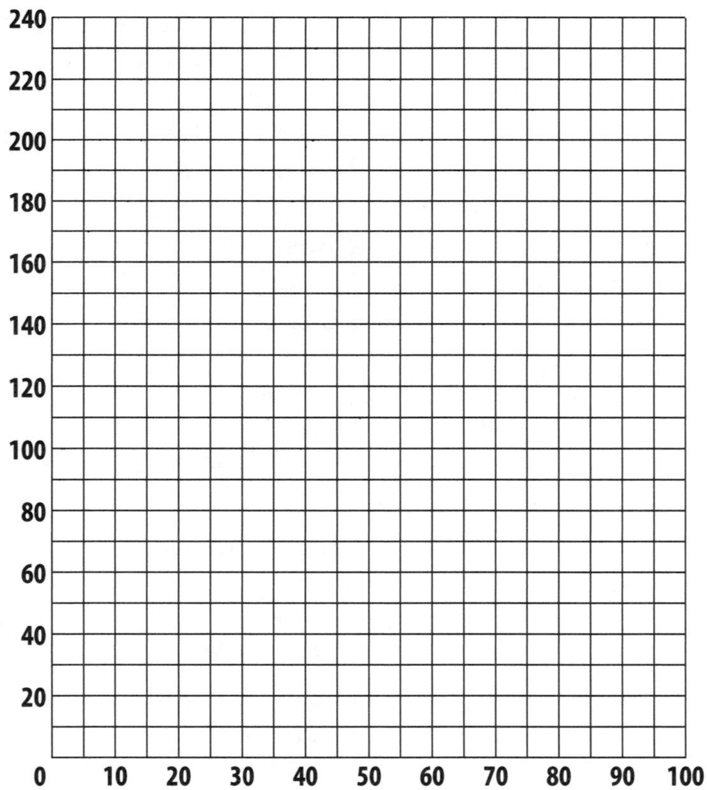

Name: ______________________________ Date: ______________

Lab 3.1.5A Freezing Point Depression

QUESTION: How does freezing point depression help make ice cream?

HYPOTHESIS: ______________________________

____

EXPERIMENT:

You will need:	• weighing paper	• vanilla extract
• graduated cylinder	• mass scale	• large ziplock bag
• cream	• small ziplock bag	• ice cubes
• sugar	• eyedropper	• rock salt

Steps:

1. Using the graduated cylinder, measure 100 mL of cream.
2. Use the weighing paper and mass scale to measure 17 g of sugar.
3. Put the cream and sugar into the small ziplock bag. Using the eyedropper, add two drops of vanilla extract and seal the bag tightly.
4. Place the small bag inside the large bag. Add 400 g of ice and 100 g of rock salt to the large bag.
5. Seal the large bag.
6. Knead the bags for 10 minutes, making sure the ice surrounds the small bag.
7. Open the large bag, and record the temperature of the ice-salt solution.

ANALYZE AND CONCLUDE:

1. What is the freezing temperature of water? ______________________________

2. What was the temperature of the ice-salt solution? ______________________________

____

3. What was the role of the salt? ______________________________

____

4. Without the added sugar, the cream would be too hard to scoop. Why? ______________

____

____

____

5. Did you make ice cream? Describe its consistency. ______________________________

____

____

____

Name: ______________________ Date: ____________

Lab 3.1.5B Boiling Point Elevation

QUESTION: How will salt affect the boiling point of water?

HYPOTHESIS: ______________________

____

EXPERIMENT:

You will need:	• salt	• 2 thermometers
• 2 small sauce pans	• spoon	• timer
• distilled water	• 2 hot plates	

Steps:

1. Fill both pans three-quarters full of distilled water.
2. Add salt to the water in the first pan, stirring constantly with the spoon, until no more salt will dissolve.
3. Adjust each hot plate to the same setting, and heat each pan on its own hot plate.
4. Place a thermometer in each pan, and start the timer.
5. Determine how long it takes for the water to boil.
6. Determine the temperature at which each solution boils.

	Boiling Time	Boiling Temperature
Saltwater Solution		
Distilled Water		

ANALYZE AND CONCLUDE:

1. How did the boiling point of the saltwater solution compare to the boiling point of the distilled water? ______________________

____

2. How does adding half as much salt affect the boiling point? ______________________

____

____

____

3. Write an experiment that compares the change in the boiling point of water when vegetable oil is added to one sample and sugar to another. ______________________

____

____

____

Name: ______________________________ Date: ______________

Contaminants WS 3.1.1A

All natural water resources contain solutes. Many minerals dissolved in groundwater are beneficial, but many are harmful. Research the following groundwater contaminants to find the sources and potential health risks.

Ground Water Contaminants

Contaminant	Sources	Potential Health Risks
Arsenic		
Barium		
Lead		
Mercury		
Silver		
Pesticides		
Chlorinated Solvents		

1. Choose two of the contaminants to research further to determine how they are removed from a water source. ______________________________

____

____

2. How does contaminated water impact people, wildlife, and the environment?

____

____

Name: ______________________ Date: ____________

Vitamins

WS 3.1.2A

Vitamins are an important part of a balanced diet. Research the following water-soluble and fat-soluble vitamins to find the sources, benefits, and consequences of too little or too much.

Vitamins

Vitamin Names (other names)	Sources	Benefits	Consequences	Solubility
Vitamin A (retinol)				
Vitamin B1 (thiamin)				
Vitamin B2 (riboflavin)				
Vitamin B3 (niacin)				
Vitamin C (ascorbic acid)				
Vitamin D (calciferol)				
Vitamin E (alpha-tocopherol)				
Vitamin K (phylloquinone)				

Name: ____________________ Date: __________

Concentration

WS 3.1.4A

Use the provided nutrition label to calculate the concentration of the following solutes in each serving. Remember that all quantities need to be measured in the same units before calculating.

1. What is the concentration of cholesterol in each serving?

2. What is the concentration of sodium?

3. Find the concentration of dietary fiber in one serving.

4. What is the solubility of each vitamin—fat-soluble or water-soluble?

5. Which of the substances in Questions 1–4 has the greatest concentration per serving?

Nutrition Facts	
Serving Size 100 g	
Amount Per Serving	
Calories 250	Calories from fat 10
	% Daily Value*
Total Fat 4%	4%
Saturated Fat 1.5%	4%
Trans Fat	
Cholesterol 50mg	28%
Sodium 150mg	15%
Total Carbohydrate 10g	3%
Dietary Fiber 5g	
Sugars 3g	
Protein 16%	
Vitamin A 1%	**Vitamin C** 3%
Calcium 2%	**Iron** 2%

*Percent Daily Values are based on a 2,000 calorie diet. Your daily values may be higher or lower depending on your calorie needs.

Practice converting between units. 1% = 10,000 ppm

6. If the concentration of ozone in the atmosphere is 0.00026%, what is it in ppm?

7. The concentration of oxygen in exhaled air is about 16%. What is the concentration in ppm?

8. Convert 14 ppm to a percentage.

9. If the concentration of carbon dioxide in the atmosphere equals 410 ppm, what is the concentration in percentage?

10. Laundry bleach has a concentration of 5.25% chlorine. How many ppm of chlorine are in the bleach?

11. How many ppm are in a 20% solution of chlorine?

Name: ______________________________ Date: ______________

Graphing

WS 3.1.4B

Using the data given, plot the solubility levels of each solute at the given temperatures. Solubility is measured in g/100 mL of water.

1. Use a different color for each solute.

	Temperature (°C)					
Solute	**0**	**20**	**40**	**60**	**80**	**100**
Sodium Chloride	**35.7**	**36**	**36.5**	**37.5**	**38**	**39.2**
Sucrose	**179**	**204**	**241**	**288**	**363**	**487**
Potassium Chloride	**28**	**33**	**38**	**44**	**50**	**55**

2. What is the independent variable? On the graph, which is its axis? ______________

____

3. What is the dependent variable? On the graph, which is its axis? ______________

____

4. What do all three solutes have in common? ______________________________

____

____

____

5. Which solute has the greatest change in solubility with change in temperature? ______________________________

____

6. What are the chemical formulas for each solute? ______________________________

____

____

____

Name: ______________________ Date: __________

Lab 3.2.1A Precipitation Reactions

QUESTION: How do some ionic compounds form precipitates?

HYPOTHESIS: ______________________

____

EXPERIMENT:

You will need:
- 7 labels
- 7 pipettes
- 0.1 M iron(III) chloride solution, $FeCl_3$
- 0.1 M lead(II) nitrate solution, $Pb(NO_3)_2$
- 0.1 M magnesium bromide solution, $MgBr_2$
- 0.1 M potassium carbonate solution, K_2CO_3
- 0.1 M potassium phosphate solution, K_3PO_4
- 0.1 M silver nitrate solution, $AgNO_3$
- 0.1 M sodium carbonate solution, Na_2CO_3
- test tube rack
- 48-well reaction plate

Steps:

1. Write the formulas of the solutions that are included in the materials list above on seven labels and attach each label to seven pipettes.
2. Fill each pipette with the corresponding solution.
3. Place the pipette bulbs in a test tube rack with the tips of the pipettes facing up.
4. Place a 48-well reaction plate on a white piece of paper. Refer to the table on the back of the page.
5. Add 5 drops of the iron(III) chloride solution, $FeCl_3$, to each of the six wells in the first column of the reaction plate.
6. Place 3 drops of the lead(II) nitrate solution, $Pb(NO_3)_2$ in well A1. Observe if a precipitate forms. Record your observations in the table.
7. Add 5 drops of each solution in the wells listed below:
   - $Pb(NO_3)_2$: B2–F2
   - $MgBr_2$: C3–F3
   - K_2CO_3: D4–F4
   - K_3PO_4: E5–F5
   - $AgNO_3$: F6
8. Continue adding 3 drops of the solutions that are listed along the left side of the reaction plate shown in the table to the appropriate wells. Observe if a precipitate forms. Record your observations in the table.

ANALYZE AND CONCLUDE:

1. What do ionic compounds do when they dissolve? ______________________

2. If ions from different compounds combine, what happens? ______________________

3. What is the ratio of solutions that formed a precipitate to solutions that did not?

____

4. Why did some solutions not react? ______________________

____

Lab 3.2.1A Precipitation Reactions

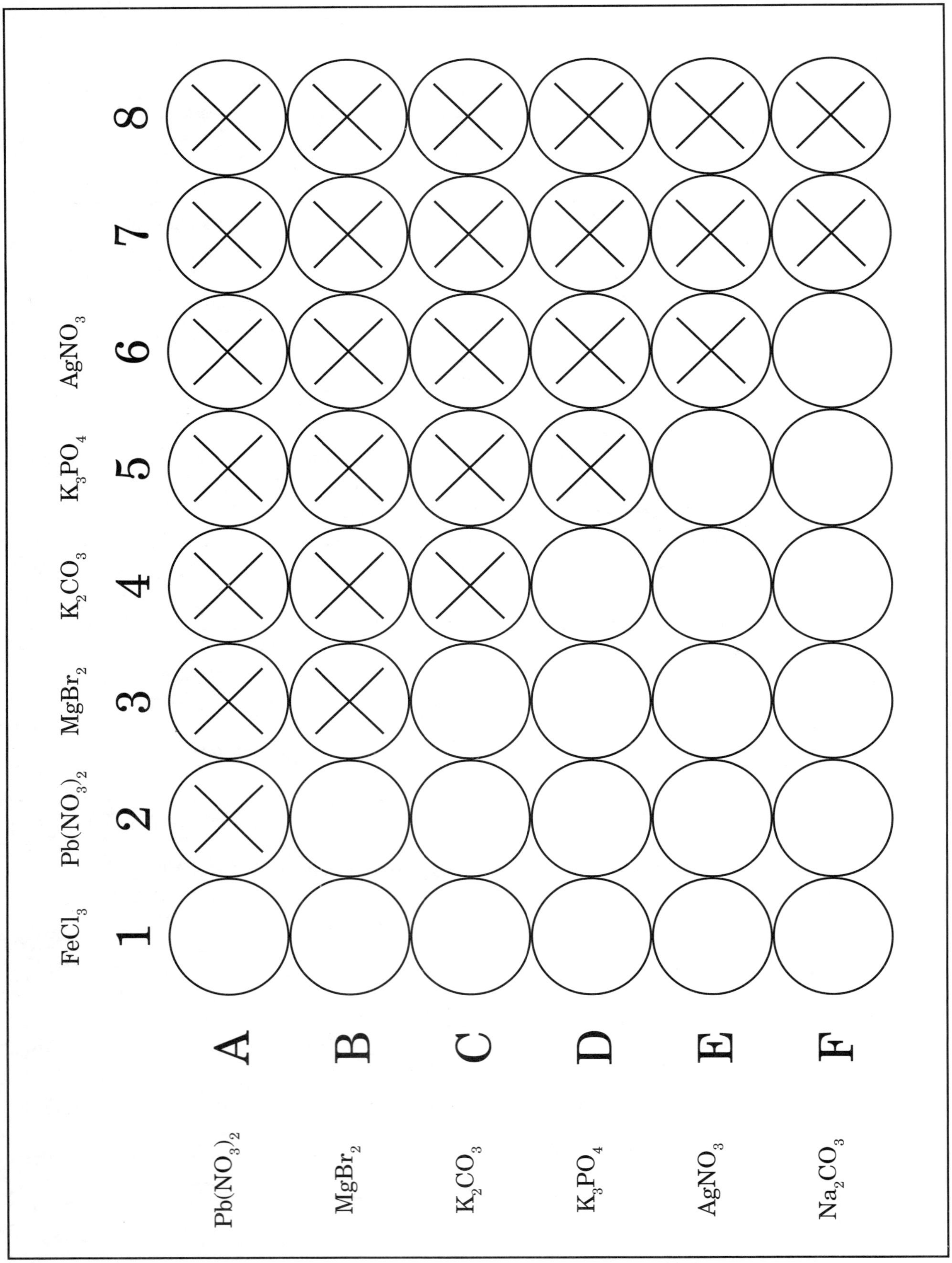

Name: ______________________________ Date: ______________

Lab 3.2.2A Milk Enzymes

QUESTION: How does the amount of enzyme affect the production of glucose?

HYPOTHESIS: ______________________________

____

EXPERIMENT:

You will need:		
• five 150 mL beakers	• stirring rod	• 7 g sodium bicarbonate
• 400 mL milk	• 6 glucose test strips	• 2 pH test strips
• 6 lactase tablets	• spatula	• graduated cylinder
• mortar and pestle	• weighing dish	• 5 mL vinegar
	• balance	

Steps:

1. Fill three beakers with approximately 80 mL of milk. One beaker will be the control.
2. Grind one lactase tablet into a fine powder with a mortar and pestle.
3. Add the powder to the second beaker of milk, and gently mix with a stirring rod.
4. Grind two lactase tablets into a fine powder with a mortar and pestle.
5. Add the powder to the third beaker of milk, and gently mix with a stirring rod.
6. Frequently stir both mixtures for about 10 minutes.
7. Using the glucose test strips, measure the amount of glucose present in the three beakers. Record your results.
   - Control group: ______
   - One lactase tablet: ______
   - Two lactase tablets: ______
8. Fill two additional beakers with approximately 80 mL of milk. The control group from Step 1 is the control group for the following steps.
9. Using a spatula and weighing dish, measure 7 g of sodium bicarbonate.
10. Pour the sodium bicarbonate into one of the beakers that is not the control group, and gently stir the solution with the stirring rod.
11. Measure the pH of the mixture with a pH test strip. Record your results.
    - Sodium bicarbonate mixture pH: ______
12. Using a graduated cylinder, measure approximately 5 mL of vinegar.
13. Pour the vinegar into the second beaker that contains milk, and gently stir with a clean stirring rod.
14. Measure the pH of the mixture with a pH test strip. Record your results.
    - Vinegar mixture pH: ______
15. Do not add any substances to the control beaker.
16. Grind one lactase tablet into a fine powder with a mortar and pestle.
17. Add the powder to the control beaker, and gently stir with a stirring rod.

Lab 3.2.2A Milk Enzymes

18. Individually grind two more lactase tablets, and add the powder from one tablet to the beaker that contains the milk and sodium bicarbonate mixture and the powder from the second tablet to the beaker that contains the milk and vinegar mixture.
19. Frequently stir the contents in the three beakers for about 10 minutes.
20. Using the glucose test strips, measure the amount of glucose present in the three beakers. Record your results.
    - Control group: ______
    - Sodium bicarbonate mixture: ______
    - Vinegar mixture: ______

ANALYZE AND CONCLUDE:

1. How did the amount of glucose compare in the first three beakers? ______________

____

____

2. Explain your results. ____

____

____

____

____

____

3. How did the amount of glucose compare in the second set of beakers? ______________

____

____

____

4. What type of substances do you think inhibit the activity of enzymes? ______________

____

____

5. Why does lactase work well in the human digestive system? ______________

____

____

____

Name: ______________________________ Date: ____________

Lab 3.2.3A Conservation of Mass

QUESTION: What happens to the mass of the reactants after a reaction?

HYPOTHESIS: ______________________________

____

____

EXPERIMENT:

You will need:		
• spatula	• 25 g sodium bicarbonate	• graduated cylinder
• balance	• weighing dish	• 25 mL vinegar
	• large ziplock plastic bag	• paper cup

Steps:

1. Using a spatula and balance, scoop approximately 25 g of sodium bicarbonate into a weighing dish.
2. Pour the sodium bicarbonate into a large ziplock plastic bag.
3. Using a graduated cylinder, measure 25 mL of vinegar.
4. Pour the vinegar into a paper cup.
5. Carefully place the paper cup inside the plastic bag without letting any of the vinegar spill out of the cup.
6. Squeeze the plastic bag to remove as much air as possible. Seal the bag.
7. Place the bag on the balance and record the initial mass.
   - Initial mass: ________
8. Without opening the bag, tip the paper cup to allow the vinegar to mix with the sodium bicarbonate.
9. Record your observations. ______________________________

   ____

   ____
10. Predict the final mass of the bag and its contents. ______________
11. Place the bag on the balance and record the final mass.
    - Final mass: ________

ANALYZE AND CONCLUDE:

1. What evidence indicates a chemical reaction occurred? ______________

____

2. Is the final mass prediction similar to the actual final mass? If not, explain why there is a difference. ______________________________

____

Lab 3.2.3A Conservation of Mass

3. How does the initial mass of the bag and its contents compare to the final mass of the bag and its contents? ______________________________

____

4. Why was the mass the same even though the amount of reactants appeared less?

____

____

5. Explain how the law of conservation of mass was maintained during this reaction.

____

____

Name: ______________________________ Date: ____________

Lab 3.2.4A Heat Reactions

QUESTION: How does heat change in endothermic and exothermic reactions?

HYPOTHESIS: ______________________________

____

EXPERIMENT:

You will need:		
• graduated cylinder	• timer	• 10 g ammonium nitrate
• two 250 mL beakers	• weighing dish	• stirring rod
• thermometer	• spatula	• 75 mL sodium hypochlorite
	• balance	• 75 mL 0.5 M sodium sulfite solution

Steps:

1. Using a graduated cylinder, measure approximately 100 mL of distilled water.
2. Pour the water into a 250 mL beaker.
3. Place a thermometer in the beaker. Do not allow the thermometer to touch the bottom or sides of the beaker. Record the temperature of the water after 2–3 minutes have elapsed.
   - Initial Temperature: ________
4. Using a weighing dish and spatula, measure 10 g of ammonium nitrate.
5. Carefully and quickly pour the ammonium nitrate into the beaker of water.
6. Using a stirring rod, gently mix the solution. Do not stir with the thermometer.
7. Measure the change in temperature every 10 seconds for 2 minutes.
8. Record your data in the table below.

Temperature Change of Water over Time													
Time (sec)	**0**	**10**	**20**	**30**	**40**	**50**	**60**	**70**	**80**	**90**	**100**	**110**	**120**
Temperature (°C)													

9. Using a clean graduated cylinder, measure approximately 75 mL of sodium hypochlorite.
10. Pour the sodium hypochlorite into a second 250 mL beaker.
11. Place a thermometer in the beaker. Record the temperature of the sodium hypochlorite after 2 minutes have elapsed.
    - Initial Sodium Hypochlorite Temperature: ________
12. Add approximately 75 mL of the sodium sulfite solution to the sodium hypochlorite.
13. Using a stirring rod, gently mix the solution. Do not stir with the thermometer.
14. Measure the change in temperature every 10 seconds for 2 minutes.
15. Record your data in the table below.

Temperature Change of Sodium Hypochlorite over Time													
Time (sec)	**0**	**10**	**20**	**30**	**40**	**50**	**60**	**70**	**80**	**90**	**100**	**110**	**120**
Temperature (°C)													

Lab 3.2.4A Heat Reactions

ANALYZE AND CONCLUDE:

1. Graph your data for the change in temperature of water in the chart below.

Temperature Change of Water over Time

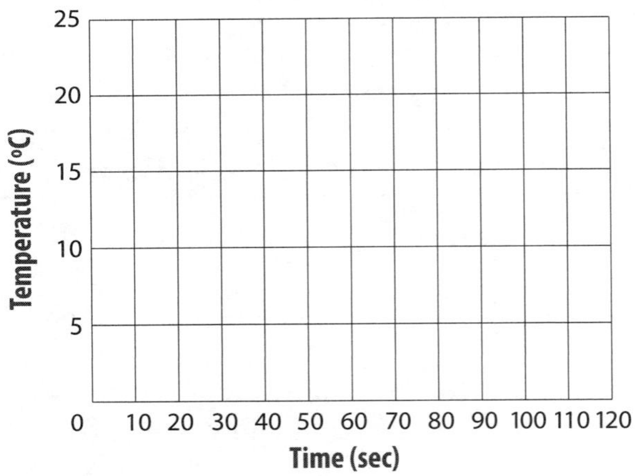

2. Graph your data for the change in temperature of sodium hypochlorite in the chart below.

Temperature Change of Sodium Hypochlorite over Time

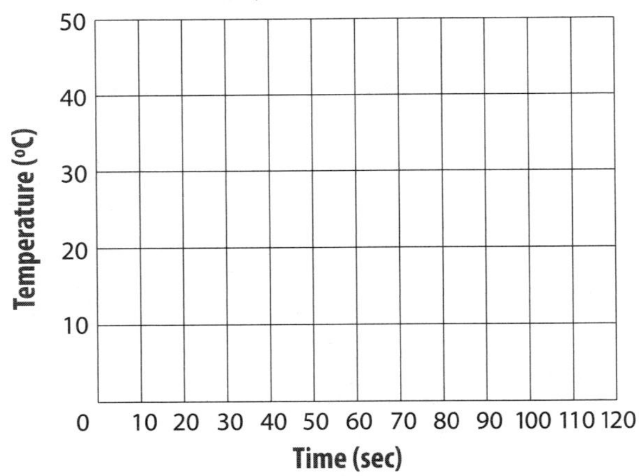

3. How did the temperature of the water change? ____________________________

____

4. Calculate the reaction rate of this reaction in °C/minute.

5. Why did this temperature change occur? ____________________________

____

____

____

6. What type of reaction occurred? ____________________________

7. How did the temperature of the sodium hypochlorite solution change? ________

____

____

8. Calculate the reaction rate in °C/minute.

9. Why did this type of temperature change occur? ____________________________

____

10. What type of reaction occurred? ____________________________

Name: ____________________ Date: __________

Lab 3.2.5A Synthesis Reaction

QUESTION: What happens to the reactants in a synthesis reaction?

HYPOTHESIS: ______________________________

____

EXPERIMENT:

You will need:	• weighing dish	• distilled water
• large test tube	• balance	• stirring rod
• test tube rack	• 1.0 g calcium oxide	• 2 red litmus test strips
• spatula	• graduated cylinder	• stopper

Steps:

1. Place a large test tube in a test tube rack.
2. Using a spatula and weighing dish, measure 1.0 g of calcium oxide.
3. Pour the calcium oxide into the large test tube.
4. Using a graduated cylinder, measure 10 mL of distilled water.
5. With a stirring rod, blot a few drops of water on a red litmus test strip. Observe and record the pH of the water.
   - Water pH: ________
6. Pour the water into the test tube.
7. Insert the stopper into the opening of the test tube, and invert the test tube multiple times for about one minute to thoroughly mix the contents.
8. Record your observations. ______________________________
9. With a stirring rod, blot a few drops of the mixture on a red litmus test strip. Observe and record the pH of the mixture.
   - Mixture pH: ________

ANALYZE AND CONCLUDE:

1. How does the pH of the water compare to the pH of the mixture? ______________

____

2. What can you conclude about the properties of the product? ______________

____

3. Write the equation for this reaction that is provided by your teacher. ______________

____

4. Using the names of each compound, explain why this reaction is a synthesis reaction. ______________________________

____

Lab 3.2.5A Synthesis Reaction

5. Why was heat produced when the water and calcium oxide were mixed? ____________

____

____

Name: ______________________ Date: __________

Lab 3.2.5B Decomposition Reaction

QUESTION: What happens during a decomposition reaction?

HYPOTHESIS: ______________________

____

EXPERIMENT:

You will need:	• 2 g potassium permanganate	• matches or flint lighter
• spatula	• large test tube	• splint
• weighing dish	• test tube clamp	• tongs
• balance	• Bunsen burner	

Steps:

1. Using a spatula and weighing dish, measure 2 g of potassium permanganate.
2. Pour the potassium permanganate into a large test tube.
3. Secure a test tube clamp around the upper portion of the test tube.
4. With the opening of the test tube pointed away from people, gently heat the base of the test tube in a Bunsen burner flame for a few minutes.
5. While holding one end of a splint with tongs, ignite the splint and quickly blow out the flame after a few seconds have passed.
6. Carefully insert the glowing tip of the splint into the opening of the test tube. Do not drop the splint into the test tube. Record your observations. ______________________
7. Allow the test tube to cool.
8. Add a small amount of water to the test tube and record the color. ______________________

ANALYZE AND CONCLUDE:

1. How did heating affect potassium permanganate? ______________________

3. Why did the splint reignite? ______________________

4. Write the equation for this reaction provided by your teacher. ______________________

5. Explain why this experiment illustrates a decomposition reaction. ______________________

6. Why did the color change? ______________________

Name: ______________________________ Date: ____________

Lab 3.2.5C Single Replacement Reaction

QUESTION: What occurs during a single replacement reaction?

HYPOTHESIS: ______________________________

____

EXPERIMENT:

You will need:	• lead(II) nitrate solution, $Pb(NO_3)_2$	• iron wire
• masking tape	• magnesium sulfate solution, $MgSO_4$	• zinc strip
• metric ruler	• zinc(II) nitrate solution, $Zn(NO_3)_2$	• magnesium ribbon
• 24-well reaction plate	• iron(II) sulfate solution, $FeSO_4$	• lead wire
• 4 pipettes	• sandpaper	• scissors

Steps:

1. Place a 13 cm piece of masking tape on the surface of the table along the upper edge of the reaction plate.
2. Write the names of each solution above the well columns according to the diagram on the following page.
3. Place a 9 cm piece of masking tape on the surface of the table along the left edge of the reaction plate.
4. Write the symbols of each element next to the well rows according to the same diagram.
5. Using a pipette, fill well A1 halfway with the lead(II) nitrate solution.
6. Continue filling the first column of wells halfway with the lead(II) nitrate solution.
7. Fill the remaining wells with the designated solutions using a clean pipette for each solution.
8. With a piece of sandpaper, remove any areas of tarnish on the 10 cm piece of iron metal wire.
9. Cut the wire into small pieces that will fit in the wells.
10. Place one piece of iron in well A1. If a reaction occurs, write the letter R in the diagram on the following page. If a reaction does not occur, write NR in the diagram on the following page.
11. Sand the pieces of zinc, magnesium, and lead to remove any tarnish.
12. Cut each metal into small pieces.
13. Add each piece of metal to the designated well and record your observations in the diagram on the following page.

Lab 3.2.5C Single Replacement Reaction

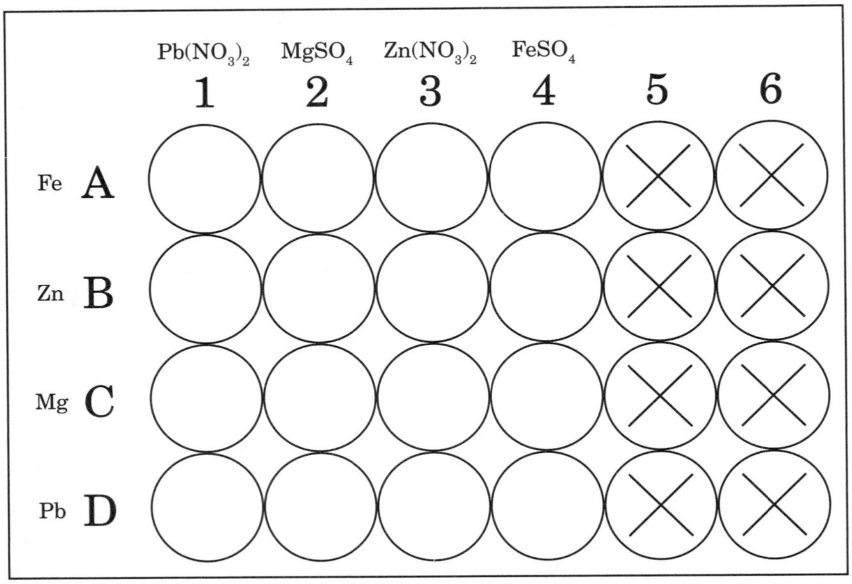

ANALYZE AND CONCLUDE:

1. Describe how you could tell if a reaction occurred. ____________________

____

2. Which element produced the most reactions? ____________________

3. What does this indicate about its activity? ____________________

4. Which element produced the least reactions? ____________________

5. What does this indicate about its activity? ____________________

6. Determine the activity series for the four elements tested, from most active to least active. ____________________

7. Refer to the metal activity series in your textbook. Does your activity series in Question 6 match the series table? ____________________

8. What specifically happened when iron reacted with lead(II) nitrate? __________

____

____

9. Write the equations for the six reactions that occurred. ____________________

____

____

____

____

Name: ______________________ Date: __________

Lab 3.2.5D Double Replacement Reaction

QUESTION: What happens during a double replacement reaction?

HYPOTHESIS: ______________________

____

EXPERIMENT:

You will need:	• test tube rack	• 5 mL aluminum chloride solution
• large test tube	• graduated cylinder	• 1 mL ammonium hydroxide solution

Steps:

1. Place a large test tube in a test tube rack.
2. Using a graduated cylinder, measure 5 mL of aluminum chloride solution.
3. Pour the aluminum chloride solution into the large test tube.
4. In a clean graduated cylinder, measure 1 mL of ammonium hydroxide solution.
5. Add the ammonium hydroxide solution to the aluminum chloride solution and record your observations. ______________________

____

ANALYZE AND CONCLUDE:

1. How do you know a reaction took place? ______________________

____

2. Write the equation for this reaction provided by your teacher.

____

3. What is the identity of the precipitate? ______________________

4. What exactly happened during this reaction? ______________________

____

____

____

5. Why is the majority of the product a liquid? ______________________

____

____

____

Name: ______________________________ Date: ______________

Reactants and Products

WS 3.2.1A

1. Circle the reactants in the chemical equations below, and circle the products in the equations using a different color.

$$2NaOH + CaBr_2 \longrightarrow Ca(OH)_2 + 2NaBr$$

$$Ti(OH)_4 \longrightarrow TiO_2 + 2H_2O$$

$$Ca(OH)_2 + 2HCl \longrightarrow CaCl_2 + 2H_2O$$

2. Two units of methanol, CH_3OH, react with three units of oxygen gas, O_2, to produce two units of carbon dioxide, CO_2, and four units of water, H_2O. Fill in the equation below for this reaction. Identify the reactants and the products.

2________ + 3________ ⟶ 2________ + 4________

3. Two units of hydrogen gas react with one unit of oxygen gas to produce two units of water. Fill in the equation below. Identify the reactants and the products. (Remember, elements that form diatomic molecules include H, N, O, and the halogens.)

2________ + ________ ⟶ 2________

4. Write the chemical equation in the space below for the following reaction: ten units of carbon dioxide and four units of water are created when one unit of decacarbon octahydride reacts with twelve units of oxygen gas.

5. Write a reaction equation for the baking of a standard vanilla cake.

Name: ______________________ Date: __________

Reaction Factors

WS 3.2.2A

Determine which reaction factor (temperature, surface area, concentration of reactant(s), catalyst, or inhibitor) is being described in the following word problems.

1. Hydrogen peroxide can slowly decompose into water and oxygen gas. To help speed up this reaction, potassium permanganate is added. After the reaction, scientists can recover the potassium permanganate because it is not affected by the reaction and can be reused later for other experiments. ______________________

2. Mill workers noticed that a pile of grain does not readily ignite. However, if grain is dispersed in the air in a grain elevator, it will explode or burn if a spark or flame is present. Because of this, mill workers adhere to very strict fire regulations.

   ____

3. Some bacterial infections spread rapidly. Scientists have created medications, such as penicillin, to slow down the spread of infection. Some of these medications prevent bacteria from being able to build cell walls that they need to survive.

   ____

4. Carbon dioxide and water do not usually react with one another at standard conditions. The molecules are not moving fast enough and do not have the adequate amount of energy to produce glucose. However, with the addition of heat, the carbon dioxide and water molecules begin to move quickly and their collision frequency increases. ______________________

5. The reaction between nitrogen dioxide and carbon monoxide occurs in two steps. One step is slow and the other step is fast. Adding more carbon monoxide to the solution does not increase the reaction rate because this reactant participates in the fast step. However, the addition of more nitrogen dioxide to the solution increases the reaction rate because this reactant participates in the slow step.

   ____

6. Describe a reaction that occurs in your daily life that is affected by at least three reaction factors. ______________________

   ____

   ____

   ____

   ____

   ____

   ____

   ____

Name: ______________________________ Date: ______________

Balancing Act

WS 3.2.3A

Balance the following equations. All of the blanks may not need to be filled in.

1. ____ $NaCl$ + ____ $F_2 \rightarrow$ ____ NaF + ____ Cl_2

2. ____ Fe + ____ $H_2SO_4 \rightarrow$ ____ $Fe_2(SO_4)_3$ + ____ H_2

3. ____ N_2 + ____ $H_2 \rightarrow$ ____ NH_3

4. ____ $SiCl_4$ + ____ $H_2O \rightarrow$ ____ SiO_2 + ____ HCl

5. ____ $AlBr_3$ + ____ $K_2SO_4 \rightarrow$ ____ KBr + ____ $Al_2(SO_4)_3$

6. ____ FeO_3 + ____ $CO \rightarrow$ ____ Fe + ____ CO_2

7. ____ CH_4 + ____ $O_2 \rightarrow$ ____ CO_2 + ____ H_2O

8. ____ $(NH_4)_3PO_4$ + ____ $Pb(NO_3)_4 \rightarrow$ ____ $Pb_3(PO_4)_4$ + ____ NH_4NO_3

9. ____ HCl + ____ $CaCO_3 \rightarrow$ ____ $CaCl_2$ + ____ H_2O + ____ CO_2

10. ____ K + ____ $MgBr \rightarrow$ ____ KBr + ____ Mg

11. ____ C_3H_8 + ____ $O_2 \rightarrow$ ____ CO_2 + ____ H_2O

12. ____ CO_2 + ____ $H_2O \rightarrow$ ____ $C_6H_{12}O_6$ + ____ O_2

13. ____ $FeCl_3$ + ____ $NaOH \rightarrow$ ____ $Fe(OH)_3$ + ____ $NaCl$

14. ____ $NaBr$ + ____ $CaF_2 \rightarrow$ ____ NaF + ____ $CaBr_2$

15. ____ HNO_3 + ____ $NaHCO_3 \rightarrow$ ____ $NaNO_3$ + ____ H_2O + ____ CO_2

16. ____ S_8 + ____ $O_2 \rightarrow$ ____ SO_3

17. ____ Na + ____ $H_2O \rightarrow$ ____ $NaOH$ + ____ H_2

18. ____ H_2SO_4 + ____ $NaNO_2 \rightarrow$ ____ HNO_2 + ____ Na_2SO_4

19. ____ $(NH_4)_2CO_3 \rightarrow$ ____ NH_3 + ____ CO_2 + ____ H_2O

20. ____ C_8H_{18} + ____ $O_2 \rightarrow$ ____ CO_2 + ____ H_2O

Name: ______________________________ Date: ______________

Reaction Mathematics

WS 3.2.4A

Original number: 10

Determine whether each statement is true or false. If the statement is true, circle the mathematical operation under True. If the statement is false, circle the mathematical operation under False. Starting with the number 10, perform each operation and write the solution in the blank space provided in the middle column. Your final answer should be equal to the original number, 10.

	True		False
1. When hydrogen and oxygen react to form water, hydrogen and oxygen are the reactants.	– 10		+ 13
2. The products are the substances that are present before a chemical reaction.	– 12		+ 4
3. The higher the reactant concentration, the faster the reaction.	× 3		× 10
4. Reactions happen faster at low temperatures.	× 5		÷ 6
5. Mass cannot be created or destroyed during a chemical reaction.	+ 3		– 9
6. A chemical equation must have an equal number of atoms of each element on the reactant side of the equation and the product side of the equation.	+ 5		× 7
7. Substances that are ground into powder react faster than cubes of the same substance.	– 10		+ 10
8. A catalyst becomes a different substance during a chemical reaction.	+ 17		+ 4
9. If heat is released during a chemical reaction, it is an exothermic reaction.	× 3		× 19
10. The following equation is balanced: $Na + Cl_2 \longrightarrow 2NaCl$.	÷ 9		÷ 6
11. An endothermic reaction absorbs heat.	+ 3		+ 21
12. Some reactions involve no change in energy.	+ 1		+ 5

Name: ______________________________ Date: ______________

Lab 3.3.1A Acid Corrosion

QUESTION: Why are some acids stronger than others?

HYPOTHESIS: ______________________________

____

EXPERIMENT:

You will need:
- 5 pre-1982 pennies
- 5 post-1983 pennies
- triangular file
- ten 50 mL beakers
- label tape
- graduated cylinder
- 50 mL dilute hydrochloric acid, HCl
- tongs
- 50 mL dilute sulfuric acid, H_2SO_4
- 50 mL acetic acid, $C_2H_4O_2$
- 50 mL carbonated water (carbonic acid), CH_2O_3
- 50 mL lemon juice (citric acid), $C_6H_8O_7$

Steps:

1. Obtain one old penny dated before 1982 and one new penny dated after 1983.
2. File three 4 mm notches in each penny with a triangular file.
3. Label one 50 mL beaker *HCl Old* and the second 50 mL beaker *HCl New*.
4. Using a graduated cylinder, carefully fill each beaker with 25 mL of dilute hydrochloric acid.
5. Grasp the old penny with a pair of tongs and gently place the penny in the appropriate beaker. Wait 2–3 minutes and record your observations in the table below.
6. Using a pair of tongs, gently place the newer penny in the appropriate beaker. Wait 2–3 minutes and record your observations in the table below.
7. Repeat Steps 1–6 using dilute sulfuric acid, acetic acid, carbonated water, and lemon juice.

Acid Name	Old Penny Reaction	Newer Penny Reaction
Hydrochloric acid		
Sulfuric acid		
Acetic acid		
Carbonated water (carbonic acid)		
Lemon juice (citric acid)		

Lab 3.3.1A Acid Corrosion

ANALYZE AND CONCLUDE:

1. Did all the pennies react with the acids? Why? ______________________

____

____

____

2. Rank the acids from most reactive to least reactive. ______________________

____

____

3. Why were some acids more reactive than others? ______________________

____

____

____

4. What property of acid did this experiment test? ______________________

Name: ____________________ Date: ____________

Lab 3.3.3A Indicators

QUESTION: When do indicators change color?

HYPOTHESIS: ____

____

EXPERIMENT:

You will need:
- 24-well reaction plate
- 6 pipettes
- unknown solution 1
- unknown solution 2
- labels
- thymolphthalein indicator
- phenolphthalein indicator
- phenol red indicator
- bromthymol blue indicator
- four 150 mL beakers

Steps:

1. Obtain a 24-well reaction plate.
2. Using a pipette, fill the wells in Column 1 half full with unknown solution 1.
3. With a clean pipette, fill the wells in Column 2 half full with unknown solution 2.
4. Write the name of each indicator on four separate labels. Attach each label to an unused pipette bulb so the name of the indicator is visible.
5. Fill each pipette bulb with the appropriate indicator, and place the pipettes with the bulbs facing up in separate beakers.
6. Add 2 drops of thymolphthalein to well A1 and well A2. Observe and record the color change in the chart below.
7. Continue adding two drops of each indicator to the following wells:
   - Phenolphthalein: B1 and B2
   - Phenol Red: C1 and C2
   - Bromthymol Blue: D1 and D2
8. Observe and record the color changes for each indicator in the chart below.

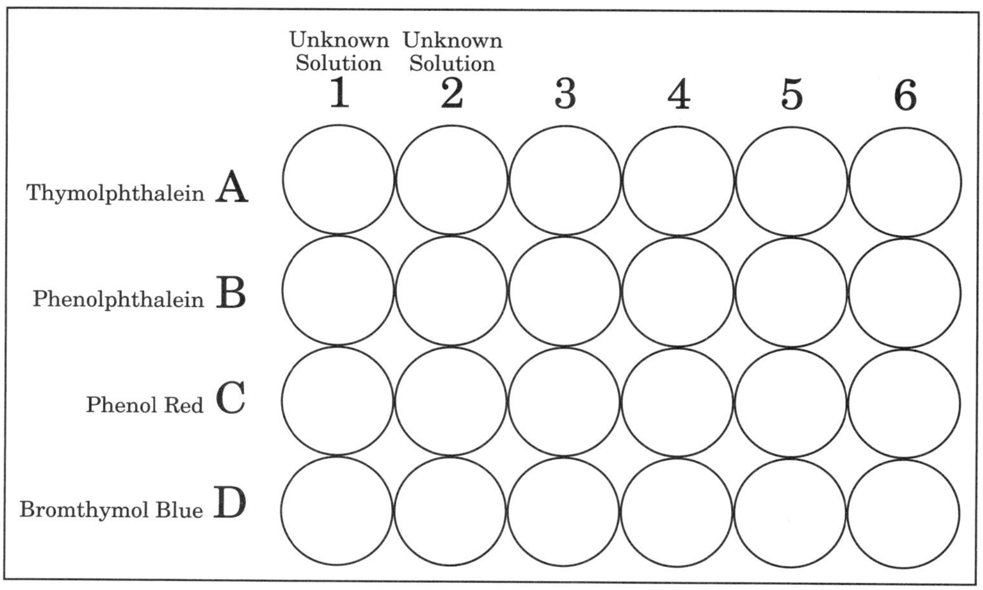

Lab 3.3.3A Indicators

ANALYZE AND CONCLUDE:

1. Research and record the pH ranges and the color changes for the four indicators tested in this laboratory experiment. ______________________________

2. Is unknown solution 1 an acid or a base? ______________________________

3. Is unknown solution 2 an acid or a base? ______________________________

4. How could the color in well A1 be changed? ______________________________

5. The predominant presence of what ion causes phenolphthalein to appear pink?

6. Could some basic solutions appear clear when thymolphthalein indicator is added? Why? ______________________________

7. What is a more precise method for measuring pH? ______________________________

Name: ______________________________ Date: ______________

Lab 3.3.4A Neutralization Reactions

QUESTION: How do bases neutralize acids?

HYPOTHESIS: ______________________________

____

EXPERIMENT:

You will need:		
• graduated cylinder	• magnetic stirring bar	• magnetic stirrer
• 150 mL distilled water	• pipette	• antacid tablet
• 250 mL Erlenmeyer flask	• hydrochloric acid	
	• universal indicator solution	

Steps:

1. Using a graduated cylinder, measure 150 mL of distilled water and pour it into the 250 mL Erlenmeyer flask.
2. Tilt the flask slightly without letting any water pour out and carefully place a magnetic stirring bar in the flask.
3. With a pipette, add 4 drops of hydrochloric acid to the water.
4. Add 16 drops of universal indicator solution to the hydrochloric acid solution. Record the color of the solution. ______________________________
5. Place the flask on the magnetic stirrer and turn on the stirrer.
6. Carefully add an antacid tablet to the flask and observe and record the color changes over time. ______________________________

   ____

ANALYZE AND CONCLUDE:

1. What type of substance is an antacid tablet? ______________________________

2. Explain what happened when the antacid tablet was added to the flask. ______________

____

____

____

____

____

3. When did the antacid neutralize the hydrochloric acid solution? ______________

____

4. Why are antacids effective in the human body? ______________________________

Name: ______________________________ Date: ____________

Lab 3.3.4B Titrations

QUESTION: When is the end point of a titration reached?

HYPOTHESIS: ______________________________

____

EXPERIMENT:

You will need:		
• 3 labels	• graduated cylinder	• orange juice
• three 125 mL Erlenmeyer flasks	• 75 mL starch-iodine solution	• juice sample 2
	• 3 pipettes	• juice sample 3

Steps:

1. Predict which juices contain the most vitamin C and which juices contain the least amount of vitamin C. Rank the juices in order from most to least vitamin C content. ______________________________

   ____
2. Write the names of the three types of juices on separate labels.
3. Attach the labels to three 125 mL Erlenmeyer flasks.
4. Using a graduated cylinder, measure 25 mL of starch-iodine solution and pour it into each flask. Record the color of the solution. ____________
5. Fill a pipette bulb with orange juice.
6. Carefully add the orange juice, one drop at a time, to the starch-iodine solution. Swirl the flask after each drop addition until the solution is clear.
7. Record the number of orange juice drops that are added to the starch-iodine solution to make the solution colorless.
8. Repeat Steps 5–7 using the other two juice samples. Record your observations in the table below.

Name of Juice	Number of Drops
Orange juice	

ANALYZE AND CONCLUDE:

1. On the next page, create a bar graph that compares the number of juice drops that were added to each solution until the end point was reached.

2. Which juice has the most vitamin C? ______________________________

3. Which juice has the least vitamin C? ______________________________

Lab 3.3.4B Titrations

4. Is your prediction of vitamin C content in agreement with these results? If not, why? ____

____

____

____

5. Is the solution neutral when the end point is reached? Explain. ______________

____

____

____

Name: ______________________________ Date: ______________

Lab 4.1.1A Elastic Force

QUESTION: How is mass related to elastic force?

HYPOTHESIS: ____

____

EXPERIMENT:

You will need:	• stand with clamp	• several different hooked masses
• rubber band	• metric ruler	

Steps:

1. Hang the rubber band on the clamp.
2. Measure the length of the rubber band and record. ______________________
3. Attach the smallest mass to the end of the rubber band. Measure the length of the rubber band, and record it in the data table.
4. Remove the mass and allow the elastic force to reshape the rubber band.
5. Repeat the above procedure with the other masses.
6. Plot the relationship between the masses and the length of the rubber band on the graph.

Mass (g)	Length of Rubber Band (cm)

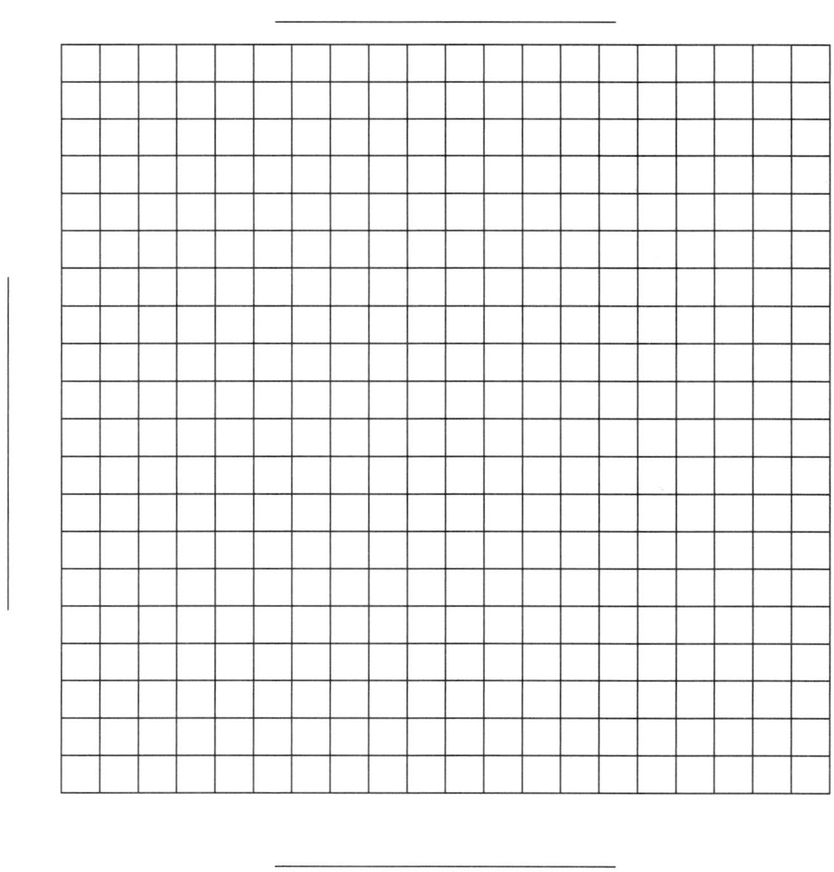

Lab 4.1.1A Elastic Force

ANALYZE AND CONCLUDE:

1. Which variable is the independent variable? ______________________

2. Which variable is the dependent variable? ______________________

3. How does mass relate to the elastic force? ______________________

____

____

____

4. How do your results compare with your hypothesis? ______________________

____

____

Name: ______________________ Date: __________

Lab 4.1.1B Friction

QUESTION: Which surface provides the greatest frictional force?

HYPOTHESIS: ______________________

____

EXPERIMENT:

You will need:	• Newton spring scale	• smooth surface
• heavy book	• carpet	• several straws
• string		

Steps:

1. Tie the string around the book and attach it to the spring scale.
2. Holding the spring scale, steadily pull the book across the carpet.
3. Record how much force, in newtons, is needed to keep the book moving. __________
4. Pull the book across the smooth surface and then the straws. Record the force necessary to keep the book moving on each surface. __________

ANALYZE AND CONCLUDE:

1. What is friction? ______________________

____

2. Which surface provided the greatest frictional force? __________

3. What evidence supports the claim? ______________________

____

____

4. How do the results compare with the hypothesis? __________

____

____

Name: ______________________ Date: __________

Lab 4.1.2A Velocity and Acceleration

QUESTION: How will velocity and acceleration change over time?

HYPOTHESIS: ______________________

____

EXPERIMENT:

You will need:	• toy car	• timer
• several books	• meterstick	
• board or Hot Wheels track, 1.5 m long		

Steps:

1. Assign group members the following jobs: one car launcher, two to four distance indicators, one timekeeper, one measurer, and one recorder.
2. Set the board or track up against a book to make a launching ramp. The top of the ramp should be 8 cm high.
3. Place the toy car at the top of the ramp, and mark a starting line in front of it.
4. Give the stopwatch to the timer and the meterstick to the measurer. Read through Steps 5–9 so each person knows what to do.
5. The distance indicators should be positioned at different locations along the ramp. (To get an idea where the distance indicators should stand, make a trial run with the car to see where it will be at one, two, three, and four seconds.)
6. When the timekeeper gives the command to start, the launcher should release the car from the top of the ramp.
7. The timekeeper should call out each second as long as the car continues to move. At the same time, the distance indicators should mark where the car is at each second.
8. After the car has stopped moving, the measurer measures from the starting line to each point that the distance indicators marked.
9. Record the results in the data table.
10. Repeat the procedure two more times and find the average distance at each point.

Time	Distance (cm) Trial 1	Distance (cm) Trial 2	Distance (cm) Trial 3	Average Distance (cm)	Average Velocity (cm/sec)
1 sec					
2 sec					
3 sec					
4 sec					
5 sec					

11. Using the calculated averages for distance, create a graph (on the following page) showing the position of the car versus time.
12. Calculate the average velocity for each second and graph the results.

Lab 4.1.2A Velocity and Acceleration

ANALYZE AND CONCLUDE:

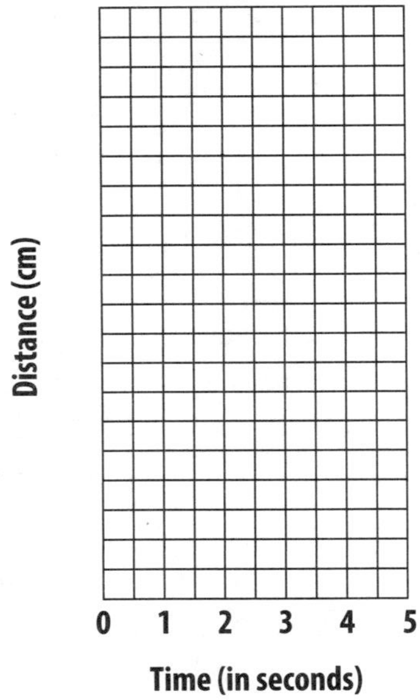

1. How far did the car travel? How many seconds did the car travel? ______________________

2. Describe the distance the car traveled in each second.

____

____

____

____

3. What happened to the velocity over time? ______________________

____

____

____

4. What is a change in velocity called? ______________________

5. How is the acceleration different at the beginning of the trip compared to the end of the trip? ______________________

____

____

6. Calculate the acceleration from start to 1 second (cm/sec^2). ______________________

7. Calculate the acceleration from the 3rd second to the 4th second. ______________________

8. What is the difference in the two accelerations? ______________________

____

____

____

____

____

9. What forces were working on the car when it was moving? ______________________

____

10. Were the forces balanced or unbalanced? What was the net force? ______________________

____

Name: ______________________ Date: ____________

Lab 4.1.2A Velocity and Acceleration continued

11. What forces were working on the car when it stopped? ____________

____

12. Were the forces balanced or unbalanced? What was the net force? ____________

____

13. If the same forces acted on the car at the beginning and the end of its travel, why did its motion change? ____________

____

____

Name: ______________________________ Date: ____________

Lab 4.1.3A Acceleration of Gravity

QUESTION: Do all objects fall with the same acceleration?

HYPOTHESIS: ______________________________

____

EXPERIMENT:

You will need:	• small rubber ball	• 5 basket-type coffee filters
• table-tennis ball	• meterstick	

Steps:

1. Assign one student to be the dropper, two students to be the watchers, and one to be the recorder.
2. The dropper should hold the table-tennis ball in one hand and the small rubber ball in the other at the height of 1 m above the floor.
3. When the watchers are ready, the dropper should release both balls at the same time. The watchers will determine if the balls hit the floor at the same time or different times. Record your observations. ______________________________

   ____
4. Repeat Steps 2–3 from a height of 2 m and record your observations.

   ____

   ____
5. Have the dropper hold one coffee filter in one hand and four nested coffee filters in the other hand.
6. Use the same procedure for the balls in Steps 2–4. Record your observations.

   ____

   ____

ANALYZE AND CONCLUDE:

1. How are objects affected by the acceleration of gravity? ______________________________

____

2. What other factors can affect the acceleration of a falling object? ______________________________

____

3. Was there a difference in the acceleration of the two balls? Why? ______________________________

____

____

____

Lab 4.1.3A Acceleration of Gravity

4. Did changing the height cause any difference in the acceleration of the two balls?

____

____

____

5. Was there a difference in the acceleration of the single coffee filter and the nested coffee filters? Why? ____

____

____

6. Did changing the height make any difference? ____

____

____

7. When do all objects fall with the same acceleration? ____

____

____

Name: ______________________ Date: __________

Lab 4.1.3B Projectile Motion

QUESTION: When launching a soft dart, what is the best angle for the greatest distance?

HYPOTHESIS: ______________________________

____

EXPERIMENT:

You will need:		
• soft dart gun	• soft darts	• meterstick
	• large protractor	• masking tape

Steps:

1. For each trial, keep the dart gun the same height above the ground.
2. Using the protractor, line up the gun with 0° and shoot.
3. Use masking tape to mark the dart's landing spot and label it *Trial 1, Angle 0°*.
4. Repeat Steps 2–3 four times, using the same angle each time.
5. Measure and record the distances of all five trials and average the results.
6. Repeat Steps 2–4, using one of the following angles each time: *15°, 30°, 45°, 60°*.

Distance Traveled (m)

Angle	Trial 1	Trial 2	Trial 3	Trial 4	Trial 5	Average
0°						
15°						
30°						
45°						
60°						

7. Create a scatterplot graph showing the distance each dart traveled at each angle. Assign a color to each angle and use that color for all distances related to it. (For example, all points for 0° are blue. All points for 15° are orange.) Show the horizontal distance the dart traveled on the y-axis and the projection angle on the x-axis. After all the points have been plotted, use black to plot the average distances and draw a line to connect the averages.

ANALYZE AND CONCLUDE:

1. What is projectile motion? ______________________________

____

____

Lab 4.1.3B Projectile Motion

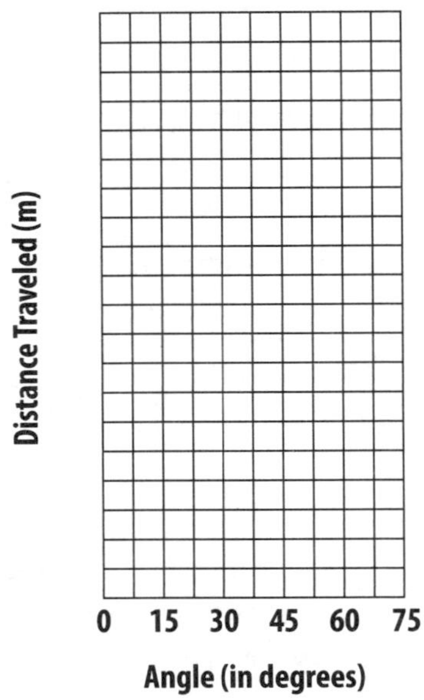

2. How do the individual distances compare with the average distances? ______________________________

____

____

3. Which angle produced the shortest distance? ______________

____

4. Which angle produced the longest distance? ______________

____

5. How do your results compare with your hypothesis? ______

____

6. Compare your results to other lab groups. How do they compare? ______________________________

____

____

____

7. Are there any distances that are not grouped with the rest? If yes, list any experiment errors that may have occurred to create this difference. ______________

____

____

____

8. Describe the shape of the line created by connecting the averages. What does the data tell you about the angle and distance? ______________________________

____

____

____

____

Name: ______________________________ Date: ______________

Lab 4.1.4A Gyroscope

QUESTION: How does a gyroscope help with stability?

HYPOTHESIS: ______________________________

____

EXPERIMENT:

You will need:	• bicycle tire with quick release hub or stunt pegs	• 2 oven mitts, used for stopping the tire

Steps:

1. Hold the bike wheel vertically by the center hub so the wheel can spin freely.
2. Before spinning the tire, try tilting the wheel left and right.
3. Try twisting the wheel around while it is vertical.
4. Hold the wheel vertically again and have a partner give the wheel a strong spin.
5. While it is spinning, try tilting the wheel left and right.
6. Try twisting the wheel around while it is vertical.
7. Hold the wheel in a horizontal position at arm's length, with one hand on the center hub.
8. Before spinning the tire, try to raise the tire to a vertical position above your head.
9. Now, with the wheel back in a horizontal position in front of you, have a partner give the wheel a strong spin.
10. Try to raise the tire to a vertical position above your head.

ANALYZE AND CONCLUDE:

1. Were you able to move the tire in Steps 2–3? ______________________________

2. How was the ability to move the tire different in Steps 5–6 from Steps 2–3? ______

____

____

3. Were you able to move the tire in Step 8? ______________________________

4. How was the ability to move the tire in Step 10 different from Step 8? __________

____

5. In this experiment, your hands and arms represented a spacecraft, and the tire represented a gyroscope. Is it likely a gyroscope could help stabilize a spacecraft? Why? ______________________________

____

____

Lab 4.1.4A Gyroscope

6. How does a gyroscope work?

7. Research how a gyroscope works, and write a paragraph to summarize how it can be used to stabilize a spacecraft.

Name: ______________________________ Date: ______________

Lab 4.1.5A Conservation of Momentum

QUESTION: How is momentum conserved in moving marbles?

HYPOTHESIS: ______________________________

____

EXPERIMENT:

You will need:		
• bendable track, such as Hot Wheels track	• meterstick	• 8 marbles
	• markers	
	• pair of 10 cm thick books	

Steps:

1. Assign the following roles: one marble releaser, several marble watchers, and one recorder.
2. Measure the length of the track and mark the center point. Also, measure and mark 5 cm intervals from each end of the track.
3. Place a book under each end of the track so the ends are 10 cm off the floor. Raising the ends will cause a slight dip in the center of the track. One group member may need to firmly hold the center of the track down without obstructing the path of the marbles, so it touches the floor.
4. Place one marble at the center point, and hold one marble 5 cm from one end of the track.
5. Assign one marble watcher to mark with a finger how far the center marble rolls. Assign another watcher to mark with a finger how far the marble released from the end rolls after it hits the marble in the center.
6. Have the marble releaser release the marble at the end of the track, and have the recorder record the observations.
7. Repeat Steps 4–6, placing three marbles at the center.
8. Repeat Steps 4–6, but change one variable at a time to test what happens. Variables may include the number of marbles in the center, the number of marbles released from the end, the point on the track from the end where the marbles are released, or the height of the end of the track.
9. Have the recorder carefully record the exact variables and the observations.

ANALYZE AND CONCLUDE:

1. In Step 6, compare velocities and masses of both marbles at the beginning. Which marble had more momentum? Why? ______________________________

____

____

____

____

Lab 4.1.5A Conservation of Momentum

2. After the first marble hit the second marble in Step 6, what happened? Why?

3. What happened when more marbles were in the center? Why?

4. Describe the most surprising observation. What happened and why?

5. If you could do the experiment again, what would you do differently?

Observations

Trial	Beginning Setup	Results
1		
2		
3		
4		

Name: ______________________ Date: __________

Lab 4.1.6A Buoyant Force

QUESTION: How does the density of an object affect buoyant force?

HYPOTHESIS: ______________________

____

EXPERIMENT:

You will need:	• Newton spring scale	• marbles
• string	• large beaker	• mass scale
• lump of modeling clay		

Steps:

1. Tie a piece of string around the clay, leaving a loop to which the spring scale can be attached.
2. Determine the weight of the clay in newtons. Record the weight. __________
3. Fill the large beaker with water. Submerge the clay in the water, but do not let it touch the sides or bottom of the beaker. Hold the spring scale in the air above the beaker.
4. How does the weight of the clay in the air compare to its weight in water?

   Explain. ______________________

   ____

   ____

   ____

5. How much buoyant force is acting on the clay? ______________________

   ____

   ____

6. Change the shape of the clay so it will float on the water.
7. Tie a string around the new clay shape, and measure its weight in the air.

   Record the weight. ______________________

   ____

8. Place the floating clay shape back in the water. Can you find the weight of the clay in water now? Why? ______________________

   ____

   ____

   ____

Lab 4.1.6A Buoyant Force

9. Add marbles to your floating shape, one at a time. Continue to do this until the clay sinks. Record how many marbles the shape could hold and still float.

____

10. What is the mass of the marbles that the clay shape can hold? ______________

ANALYZE AND CONCLUDE:

1. What is buoyant force? ______________

____

2. How does the density of the object affect buoyant force? ______________

____

____

____

3. How did the density of the clay change even when the mass stayed the same? ______

____

____

____

____

4. What similarities does your floatable design have with the others in the class?

____

____

____

____

5. Which designs held the most marbles and which held the least? Why? ______________

____

____

____

____

____

____

____

Name: ______________________________ Date: ______________

Lab 4.1.6B Density and Buoyancy

QUESTION: How does the density of a liquid affect buoyant force?

HYPOTHESIS: ____

____

EXPERIMENT:

You will need:
- three 100 mL beakers
- mass scale
- sugar solution
- oil
- cork
- 100 mL graduated cylinder
- rubber stopper

Steps:

1. Place an empty beaker on the mass scale, and record its mass. ______________
2. Take the beaker off the scale, and fill it with 50 mL of water.
3. Place the filled beaker on the scale, and record its mass. Calculate the mass and density of the fluid in the beaker.
4. Repeat Steps 1–3 with the other two fluids.

Fluid	Mass of Beaker	Mass of Beaker and Fluid	Mass of Fluid	Density of Fluid (g/mL)
Water				
Sugar solution				
Oil				

5. Place the cork on the scale, and record its mass. ______________
6. Add 70 mL of water to the graduated cylinder. Carefully place the cork in the graduated cylinder, pushing it down until it is just below the surface of the water.

   Record the volume of the cork in mL. Calculate its density. ______________

   ____

7. Repeat Steps 5–6 for the rubber stopper. ______________

   ____

8. Predict whether the cork and the rubber stopper will float, be suspended, or sink in each of the fluids. Record your predictions in the chart below.
9. Place the cork first in the water, then in the sugar solution, and then in the oil to test your prediction. Record the results.
10. Repeat Step 9 using the rubber stopper.

Object	Water Prediction	Water Actual	Sugar Solution Prediction	Sugar Solution Actual	Oil Prediction	Oil Actual
Cork						
Rubber stopper						

Lab 4.1.6B Density and Buoyancy

ANALYZE AND CONCLUDE:

1. Compare the density of the objects with their behavior in the different fluids.

2. What is the relationship of fluid density to buoyancy?

3. What are some other objects that could be tested in the different fluids?

4. What fluids might have a greater buoyant force than the sugar solution?

Name: ______________________________ Date: ______________

Lab 4.1.6C Fluids in a Closed System

QUESTION: How do air and water behave in a closed system?

HYPOTHESIS: ______________________________

____

EXPERIMENT:

You will need:		
• safety goggles	• three 4 cm lengths of 6 mm tubing	• T or Y air tube connector
• two 20 mL syringes	• 40 cm length of 6 mm tubing	• two 5 mL syringes
	• straight connector	

Steps:

Pneumatic System

1. Put on safety goggles.
2. Push the plungers of the two 20 mL syringes all the way in. Connect the two syringes with a 4 cm piece of tubing.
3. Try pulling the plungers back. Record observations. ______________________________

   ____
4. Disconnect the tubing. Pull one plunger all the way out. Reconnect the syringes with the tubing.
5. Push the plunger in. Record your observations. ______________________________

   ____
6. Connect the two syringes with the 40 cm length of tubing and repeat Steps 3–4.

   Record observations. ______________________________

   ____
7. Use a straight connector between two 4 cm pieces of tubing to connect the

   syringes. Repeat Steps 3–4. Record. ______________________________

   ____

   ____
8. Use the T or Y connector and three 4 cm pieces of tubing to connect the two 5 mL syringes (with plungers depressed) with one 20 mL syringe (with plunger pulled out to the 10 mL mark).
9. Record the volume of air that is in the large syringe. ______________________________
10. Push in the plunger of the 20 mL syringe. Record observations. ______________________________

    ____

    ____

Lab 4.1.6C Fluids in a Closed System

Hydraulic System

11. Push in the plungers of the 5 mL syringes. Compare the force needed to push in the plungers of the 20 mL syringe with the force needed for the 5 mL syringes.

12. Predict what will happen when you use water instead of air as the fluid in the system.

13. Repeat Steps 1–10, filling the syringes with water to make a hydraulic system. Make sure there are no air bubbles.
14. Record the differences you noticed between the pneumatic, or air, system and the hydraulic, or water, system.

ANALYZE AND CONCLUDE:

1. Which fluid compresses more easily, air or water?

2. Which fluid was harder to move through the system?

3. How do your predictions compare with your results?

4. Which system transfers force better? How does friction factor into the compression within the two systems?

Name: ______________________ Date: __________

Lab 4.1.7A Paper Airplanes

QUESTION: How does added drag affect the flight of a plane?

HYPOTHESIS: ______________________

____

EXPERIMENT:

You will need:	• masking tape	• scissors
• copy paper or slightly heavier paper	• small paper clips	• cellophane tape
• meterstick		

Steps:

1. Label the plane below to show the direction of the four forces that are exerted on a plane during flight—thrust, drag, lift, weight.

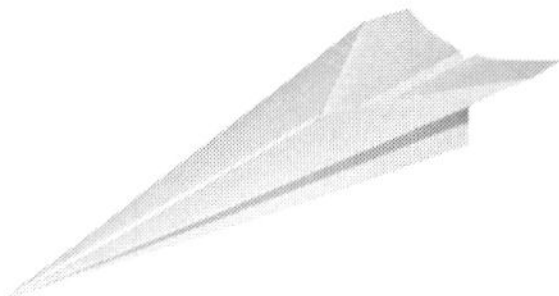

2. Build three identical paper airplanes, using **BLM 4.1.7A Airplane Instructions.**
3. Place a piece of masking tape on the floor to mark the launch line. Make note of where you are holding the plane when you launch it. ______________________
4. Launch the first plane. Measure the distance it flew and record. Add any important notes about the flight path and pattern. For example: *It flew upside down*; *it rolled as it flew*; *it went straight up and came back down*; *the nose was damaged so I had to straighten it*.
5. Launch the plane four more times and record the results.
6. After five flights with Plane 1, calculate the average distance.
7. Change the plane to increase its drag. Use scissors to cut slits 2.5 cm long where the wings meet at the middle.
8. Fold up the 2.5 cm cut section on both wings so these sections are at a 90° angle to the rest of the wing.

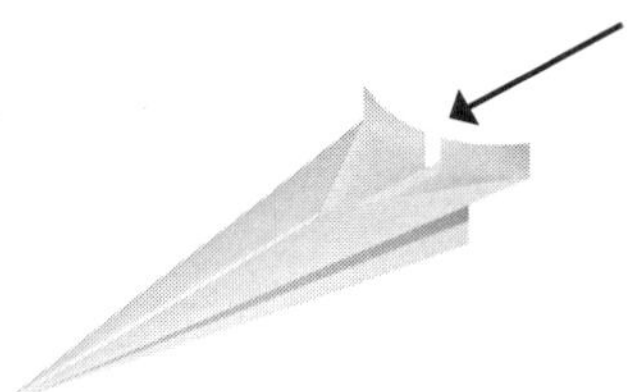

9. Launch Plane 1 with the added drag. Record the distance it flew and make note of its flight path and pattern.
10. Launch Plane 1 with the added drag four more times. Record the distances.
11. Repeat Steps 4–10 for Plane 2 and Plane 3.
12. Create a bar graph comparing the average distance for each plane with and without drag.
13. What will the *x*-axis label be? ______________________
14. What will the *y*-axis label be? ______________________

Lab 4.1.7A Paper Airplanes

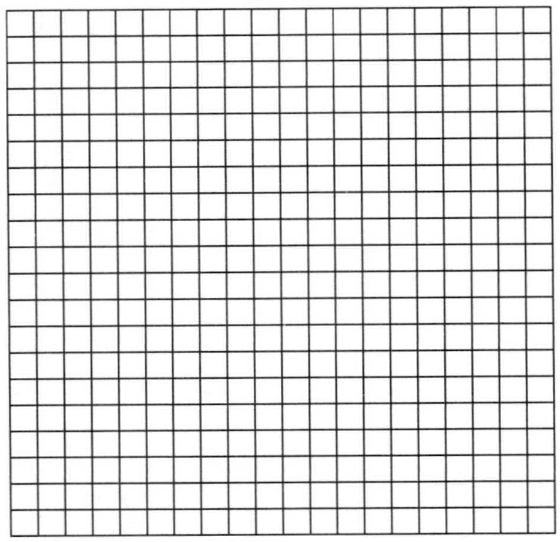

ANALYZE AND CONCLUDE:

1. Which plane averaged the shortest flight? Why?

2. Which plane averaged the longest flight? Why?

3. What other variables could be tested?

4. Write an experiment to test one variable.

Name: ____________________ Date: __________

Lab 4.1.7A Paper Airplanes continued

Paper Plane	Flight 1	Flight 2	Flight 3	Flight 4	Flight 5	Average	Notes
Plane 1							
Plane 1 with added drag							
Plane 2							
Plane 2 with added drag							
Plane 3							
Plane 3 with added drag							

Name: ______________________________ Date: ______________

Calculations

WS 4.1.1A

Solve the following problems. Convert units when necessary. Show your work.
Remember: $1\ N = 1\ kg \times m/sec^2$

1. If Jamal has a box with a mass of 2 kg, what is its weight?

2. Sheila has a pet whose mass is 631 g. How much does the pet weigh?

3. Two people carry packages to the post office. One package has a mass of 3.5 kg. The other package has a mass of 7.1 kg. What is the total weight of both packages?

4. Phillip can lift 80 N of weight. If he has 10 boxes to move, each with a mass of 4.5 kg, how many can he move at one time? How many trips will he have to make?

5. Yuki has a backpack full of books. The mass of the backpack is 200 g. Each of the 5 books has a mass of 500 g. What is the weight of the backpack full of books?

6. Consider two objects that are 8 m apart. If the two objects are moved to 1 m apart, how has the gravitational force changed?

7. Two rocks are 3 m apart. Calvin and Herman move one of the rocks and now they are 15 m apart. Is the gravitational force greater or smaller? What is the change in force between the two rocks?

8. Two tables are placed near each other. The first table has a mass of 12 kg. The second table has a mass of 24 kg. If the first table is replaced by a new table with a mass of 48 kg, what change is there to the gravitational force?

Calculations WS 4.1.1A

9. Consider two objects where the first object is 35 kg and the second object is 20 kg. Change the first object for a 70 kg object and the second for a 100 kg. What is the change in gravitational force?

10. Object 1 has a mass of 56 kg and Object 2 has a mass of 80 kg. If Object 1 is replaced with a new object that has a mass of 8 kg, how has the gravitational force changed?

11. Two trucks are parked 1 m apart. The first truck has a mass of 3,000 kg. The second truck has a mass of 1,200 kg. If the trucks are moved to 4 m apart and the first truck is replaced with a 6,000 kg truck, how has the gravitational force changed?

12. Jemero places two blocks 24 m apart. The first block is 18 kg and the second block is 48 kg. He decides the first block is too small, and he replaces it with a 54 kg block. The second one is too big so he replaces it with a 6 kg block. Then he realizes they are too far apart. He moves the two blocks within 3 m of each other. What is the final change in gravitational force?

Name: ______________________________ Date: ______________

Force and Motion

WS 4.1.2A

Answer the following questions. Show your work and convert when necessary.

1. Label the following as *speed*, *velocity*, *acceleration*, or *other*:

- 45 cm ____________________
- 84 kph ____________________
- 9.8 m/sec^2 downward ____________________
- 201 L ____________________
- 79 m/sec west ____________________
- –330 cm/sec^2 east ____________________
- 56 kph up the hill ____________________
- 124 km ____________________

2. If a bus is traveling 45 kph to the east, and a car is traveling away from the bus at 88 kph to the west, what is the relative velocity?

3. Joseph is riding his bike at 10 kph to the south. Maria is about 5 km behind him and traveling at 15 kph to the south. If they maintain constant velocities, will Maria catch up with Joseph? What is their relative velocity?

4. If Cozette pulls a wagon with 42 N of force and the frictional force is 40 N, is the wagon moving? Is the force balanced or unbalanced?

5. A train traveled 812 km in 14 hours. What was its average speed?

6. If you walk at a constant speed of 3 m/sec for 15 minutes, how far will you go?

7. Which speed is faster, 3.2 m/sec or 8 kph? Explain.

8. A plane traveled 2,230 km west in 5 hours. What was its average velocity?

9. David is skating at 4 m/sec when he notices a friend standing nearby. He increases his speed. Ten seconds later his speed is 6 m/sec. What is his acceleration?

10. Fiona is driving her car at 20 m/sec when a ball rolls into the street 5 m in front of her. She hits the brakes and stops 2 sec later, just 1 m in front of the ball. What is her acceleration?

11. If a car takes 6 sec to increase its velocity from 4.3 kph east to 40 kph east, what is its acceleration in m/sec^2?

12. Maria left her house bicycling at 10 kph north. Her acceleration was 30 km/hr^2 north. How fast was she bicycling when she arrived at school 30 min later?

Name: ______________________ Date: __________

Force Problems

WS 4.1.3A

Solve the following problems. Include units in your answers and convert when necessary.

1. How much acceleration is produced by a net force of 16 N exerted on an object of 4 kg mass?

2. To produce an acceleration of 30 m/sec^2 on a volleyball with a mass of 500 g, what net force would be needed?

3. A 12 N net force is applied to a 3 kg rock. What will the acceleration of the rock be?

4. A 12 N net force is applied to a 6 kg rock. What will the acceleration of the rock be?

5. A 550 g ball is kicked horizontally at 25 m/sec at an acceleration of 1,000 m/sec^2. What was the average force on the ball?

6. Sam pushes a car with 600 N of force. His friend pulls the car with 800 N of force in the same direction Sam is pushing, and the frictional force of the car is 1,250 N. What is the net force? Is the car moving?

7. Lola pulls a 20 kg cart with 94 N of force. The cart exerts 48 N of friction. What is the resulting acceleration?

8. A baby drops a 0.45 kg toy, which gravity causes to accelerate at 9.8 m/sec^2. What force does gravity exert on the toy?

Force Problems WS 4.1.3A

9. A 1,000 kg car has run out of gas. The owner is pushing it to the side of the road at an acceleration of 0.05 m/sec^2. What is the net force applied to the car?

10. An empty 25,000 kg truck has an engine that allows it to accelerate at 1.5 m/sec^2. At what rate can the truck accelerate after it has been loaded with 1,500 kg of cargo?

11. A can is dropped from the roof of a building. Ignoring air resistance, if the can falls in 1.3 seconds, how many meters tall is the building?

12. A sandbag falls from a hot air balloon. Ignoring air resistance, if it takes 8 seconds for the bag to fall to the ground, what is the altitude of the balloon?

Name: ______________________________ Date: ______________

Lab 4.2.1A Work and Energy

QUESTION: What determines the amount of energy needed to do work?

HYPOTHESIS: ______________________________

____

EXPERIMENT:

You will need:		
• string	• several large books	• meterstick
	• Newton spring scale	

Steps:

1. Tie a string around the first book so it can be attached to a spring scale.
2. Attach the book to the scale and record its weight in newtons.
3. Choose a distance to lift the book. Record your data.
4. Lift the book that distance.
5. Calculate the energy it took to lift the book.
6. Repeat Steps 1–5 with four different books and lift them the same distance each time.

Book	Weight (N)	Distance (m)	Energy (J); ($W = F \times d$)

7. Choose the heaviest book and lift it five different distances.
8. Record the distance and calculate the energy required in each case.

Book	Weight (N)	Distance (m)	Energy (J); ($W = F \times d$)

ANALYZE AND CONCLUDE:

1. Which book required the most energy to lift? ______________

2. What type of energy did it have when it was being lifted? ______________

____

Lab 4.2.1A Work and Energy

3. If the book is left at the height it was lifted, what type of energy would it have? How would it compare to the energy required to lift it? ______________________

____

____

4. In Steps 7–8, which distance required the greatest amount of energy? Why?

____

____

____

5. Write a formula to calculate the energy needed to lift all five books a certain distance. Use the formula to calculate the energy needed to lift all five books at the same time. ______________________________________

____

____

____

Name: ______________________ Date: __________

Lab 4.2.2A Transfer of Energy

QUESTION: How is energy transferred?

HYPOTHESIS: ______________________

____

EXPERIMENT:

You will need:		
• heavy mass	• cup hook	• small cardboard box
• 2 m piece of string	• use of a door frame	• meterstick
	• masking tape	• mass scale

Steps:

1. Tie the mass to one end of the string.
2. Secure the cup hook to the top of a door frame. Be sure that a pathway 2 m long on each side of the doorway is clear so when the pendulum swings it will not hit anything.
3. Tie the other end of the string to the cup hook so the bob, or heavy mass, has at least a 3 cm clearance from the floor when the string hangs straight down.
4. Place a piece of masking tape on the floor directly under the pendulum.
5. Lift the pendulum bob out to one side so the string is taut and the bob is 1 m from the floor.
6. Place the empty box under the door frame so the edge of the box is lined up with the piece of tape. The collision of the pendulum and the box must be head-on.
7. Release the pendulum, and watch what happens when the pendulum hits the box. Record your observations about what happened to the pendulum and the distance the box moved.
8. Repeat Steps 6–7, changing the height of the pendulum before release to 1.5 m.
9. Repeat Steps 6–7, changing the height of the pendulum before release to 0.5 m.

Initial Height of Pendulum	Distance the Box Moved	Notes

ANALYZE AND CONCLUDE:

1. What type of energy did the pendulum have when held out to the side before releasing? ______________________

2. What type of energy did the pendulum have when it hit the box? ______________________

____

Lab 4.2.2A Transfer of Energy

3. What happened to the pendulum's energy when it hit the box? ______________________

____

____

4. What is the evidence, or how do you know? ______________________________

____

____

5. What is the relationship between the amount of work done on the box by the pendulum and the different heights from which it was dropped? ________________

____

____

____

6. What force was acting on the box to prevent it from moving indefinitely? __________

____

7. Identify instances of Newton's laws of motion. ___________________________

____

____

____

____

____

____

8. Find the mass of the bob and calculate the gravitational potential energy of the bob at all three heights.

Name: ____________________ Date: ____________

Lab 4.2.4A Leg Power

QUESTION: What is the power of my leg muscles?

HYPOTHESIS: ____________________

____

EXPERIMENT:

You will need:		
• flight of stairs	• meterstick	• timer
	• bathroom scale	

Steps:

1. Measure the vertical height of a flight of stairs in meters. ____________
2. Measure your mass in kilograms. (1 pound = 0.45 kilograms) ____________
3. Calculate the force of gravity acting on your mass. ($m \times 9.8$ m/sec^2 = *Force* in newtons) ____________
4. Calculate the work in joules that is done in climbing the stairs. ($W = F/d$) ____________
5. Climb the stairs quickly. Time how long it takes you to do this. Record the time in seconds. ____________
6. Calculate your leg power in watts. ($P = W/t$) ____________

ANALYZE AND CONCLUDE:

1. What is the difference between your leg power and the average walking power of 230 W? ____________

2. How does your leg power compare to an Olympic sprinter with a leg power of 1,700 W? ____________

3. Calculate the average leg power of your class. ____________

4. How would your leg power change if you traveled down the stairs? ____________

____

5. Test your hypothesis and record your results. ____________

____

6. Write a procedure that would test your average power output going up and down the stairs for one minute. ____________

____

____

Name: ______________________________ Date: ______________

Lab 4.2.5A Reducing Friction

QUESTION: How can friction be reduced?

HYPOTHESIS: ______________________________

____

EXPERIMENT:

You will need:
- 18 cm length of string
- pens or pencils
- shoe box
- brick or other heavy object
- Newton spring scale

Steps:
1. Tie the string into a loop.
2. Use the point of a pen or pencil to poke a hole through the center of the end of the shoe box.
3. Thread the loop of string through the hole so part of the loop extends out of the hole and the rest remains inside.
4. Insert the pen or pencil through the part of the loop that is still inside the box. Pull the loop tight from the outside so the pencil holds the loop firmly in place.
5. Place the brick or other heavy object inside the shoe box.
6. Attach the spring scale to the loop of string that is on the outside of the box.
7. Place the shoe box on a smooth surface, and use the spring scale to pull the box with a horizontal force across the surface.
8. Record both the force needed to start the box moving and the force needed to keep the box moving. ______________________________
9. Repeat Steps 7–8 several times, changing one of the variables each time before you start pulling the box. Make some changes that will increase friction and other changes that will decrease it. Record the changes and the corresponding results.

Trial	Change Made	Starting Force	Sliding Force

Lab 4.2.5A Reducing Friction

ANALYZE AND CONCLUDE:

1. Which was greater, the starting friction or the sliding friction? What is the evidence, or how do you know? ______________________________

____

____

2. What changes resulted in the greatest friction? ______________________________

____

____

3. What changes resulted in the least friction? ______________________________

____

____

Name: ______________________________ Date: ______________

Lab 4.2.5B Hovercrafts

QUESTION: What does a hovercraft need in order to move?

HYPOTHESIS: ______________________________

____

EXPERIMENT:

You will need:		
• pop-top lid from plastic drinking bottle	• compact disc (CD)	• timer
	• craft glue	
	• 12" balloons	

Steps:

1. Glue the base of the pop-top lid to the CD so it covers the hole in the center of the CD.
2. Allow the glue to dry completely.
3. Make sure the pop-top lid is closed.
4. Inflate the balloon as large as you can without popping it. Count the number of breaths to fill it and record the number. ______________
5. Pinch the neck so no air can escape.
6. Stretch the neck of the balloon over the pop-top lid; be careful not to let any air escape. Center the balloon's opening above the pop-top lid opening.
7. Place the hovercraft on a flat surface.
8. Start the timer, open the pop-top lid fully, and push the hovercraft. Stop the timer when the hovercraft stops hovering.
9. Record the time in seconds that the hovercraft hovered above the surface.
10. Repeat the procedure four more times and record the results. Find the average time the hovercraft hovered.
11. Follow Steps 3–10, inflating the balloon to a medium size and then inflating the balloon to a small size.
12. Create a bar graph that shows the average hover time for each size balloon.

Hovering Time (sec)

Trial	Large Balloon	Medium Balloon	Small Balloon
1			
2			
3			
4			
5			
Average			

Lab 4.2.5B Hovercrafts

ANALYZE AND CONCLUDE:

1. What is the purpose of the air released from the balloon? ______________________

____

2. Which size balloon had the highest average hover time? ______________________

____

3. Which size balloon had the highest hover time? If the answer is not the same as the answer above, why not? ______________________

____

____

____

4. How do the results compare with the hypothesis? ______________________

____

____

5. What variables may have given incorrect results? ______________________

____

____

____

____

Name: ____________________ Date: ____________

Work Problems

WS 4.2.1A

Solve the following problems. Show all computations. Convert units when necessary and include the units with the answer.

1. How much work does a boy whose mass is 52 kg do in order to climb up a vertical ladder that is 6.2 m high?

2. How much work is done to push a stalled car 150 m by pushing on its bumper with a force of 200 newtons?

3. If you lift a 1 N apple 0.25 m to your mouth, how much energy do you exert?

4. To stop a car, an average braking force of 700 N is exerted over a distance of 30 m. How much work is done?

5. You push a cart 94 ft with a constant 42 N of force. How much work did you do?

6. You do 3 J of work to slide a 2 N container of sugar cookies across a table to your friend. How far did the cookies slide across the table?

7. You are pushing a 50 N empty cart down the hall when your friend jumps on the cart. You do 8,250 J of work to push him 15 m. What is the weight of your friend in newtons?

8. A child with a mass of 38 kg jumps off a high dive that is 6 m in the air. A child with a mass of 55 kg jumps off the low dive, which is only 2 m in the air. On which child was gravity working more? How much more?

Name: ______________________________ Date: ____________

Change in Potential Energy

WS 4.2.2A

Calculate the gravitational potential energy for a 40 kg and an 80 kg stone that are placed at varying heights. Make a line graph demonstrating the change in energy related to the change in height.

40 kg	
Height	**Energy in Joules/kilojoules**
5 m	
6 m	
7 m	
8 m	
9 m	
10 m	

80 kg	
Height	**Energy in Joules/kilojoules**
5 m	
6 m	
7 m	
8 m	
9 m	
10 m	

1. What is the independent variable? ______________________

____

2. What is the dependent variable? ______________________

____

3. How is the gravitational potential energy affected by change in height? ______________________

____

____

____

____

____

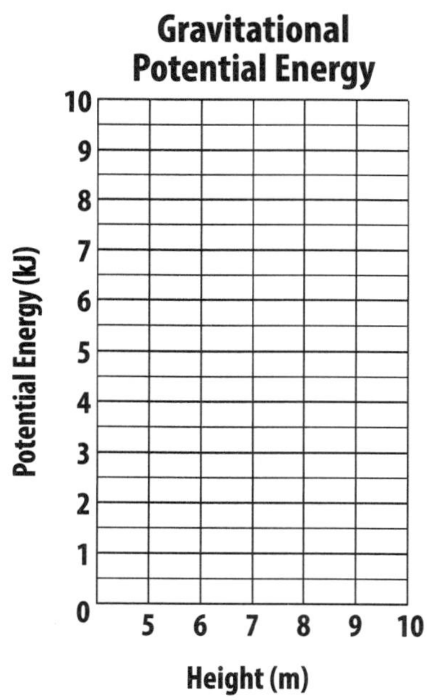

4. How does the change in energy of the 40 kg stone compare with the change in energy of the 80 kg stone? ______________________

____

____

5. What would the line look like for a 200 kg stone dropped from the same heights?

____

____

Name: ________________________________ Date: ______________

Energy In, Energy Out

WS 4.2.2B

Horatio wanted to see how much energy he consumed and how much of that energy he converted to kinetic energy, so he recorded his food intake and activities for a 24-hour period.

Food	Calories	Activity	Calories Burned per Hour for That Activity	Time Spent	Total Calories Burned
2 eggs	160	walking between classes	60	30 min	
granola bar	120	sitting in class	70	6 hr	
3 slices of cheese pizza	750	doing homework	80	2 hr	
		riding bike	470	30 min	
		walking	60	1 hr	
		sitting at computer	70	3 hr	
cheeseburger	500	soccer practice	300	2 hrs	
french fries	400	watching TV	60	2 hrs	
lemonade	100	sleeping	50	8 hrs	
Total Calories Consumed				**Total Calories Burned**	

1. How many Calories did Horatio consume? ______________________

2. If 1 Calorie = 4.184 kJ, how many kJ of energy did Horatio consume?

____

____

3. How many kJ of energy did he burn? ______________________

____

4. What type of energy did he consume? ______________________

5. What types of energy was the food changed into? ______________________

____

Name: ____________________ Date: __________

Energy Use

WS 4.2.3A

Complete the following information about the amount of energy used in your house in the last 12 months. You can find most of this information on utility bills. If your bill covers two months, divide the amount by 2; then, record that amount for both months.

1. Which utility company supplies your electricity? ____________________

____

2. What is the unit rate for electricity? (Ex. $/kilowatt-hour) ____________________

3. What energy source or sources does your electric company use to make the electricity it provides to your home? ____________________

____

4. What utility company supplies your natural gas? ____________________

5. What is the unit rate for gas? ____________________

6. Where does your gas company get its natural gas? Where are the wells of your provider? ____________________

____

7. What type of fuel does your furnace use? ____________________

8. Record your gas and electricity use for the past 12 months.

Electricity

Month	Jan	Feb	Mar	Apr	May	Jun	Jul	Aug	Sep	Oct	Nov	Dec
Amount Used												

Gas

Month	Jan	Feb	Mar	Apr	May	Jun	Jul	Aug	Sep	Oct	Nov	Dec
Amount Used												

9. Create a line graph showing both electricity and gas usage for the year.

10. What is the independent variable and what axis will it be on? ____________________

____

____

11. What is the dependent variable and what axis will it be on? ____________________

____

____

Name: ______________________________ Date: ______________

Power at Home

WS 4.2.4A

Find the power ratings of electrical devices used in your home. Estimate the number of hours per day that each device is on. Use the data chart to record the power needed for each appliance, the estimated use in hours per day, and the estimated energy in kilowatts.

Appliance	Power Rating	Estimated Hours/Day	Energy Use in Watts	Energy Use in Kilowatts

1. Find the sum of kilowatt hours used in a day and compare it to a utility bill, if the bill gives a breakdown by day. ______________
2. Which appliance could be used less often to save energy? ______________
3. What other changes could be made to use less energy? ______________

____

____

____

Name: ______________________________ Date: ______________

Review

WS 4.2.5A

Choose the letter of the best example for each term.

1. Energy
ⓐ the ability to hit a hockey puck and send it flying across the ice
ⓑ the ability to stand for a long period of time

2. Kinetic energy
ⓐ the energy in a Frisbee that is sailing through the air
ⓑ the energy in a Frisbee that you are about to throw

3. Potential energy
ⓐ the heat energy in fire
ⓑ the chemical energy in a battery

4. Law of conservation
ⓐ You use the chemical energy from the food that you ate for lunch to play basketball.
ⓑ You fell asleep after lunch, so the energy from your food was wasted.

5. Nonrenewable resources
ⓐ A house is heated by natural gas.
ⓑ A house is heated by the sun's rays.

6. Renewable resources
ⓐ Wind power is used to generate electricity for a town.
ⓑ A factory burns coal to produce heat.

7. Effort force
ⓐ You walk to your piano lesson.
ⓑ You bike to your piano lesson.

8. Ideal mechanical advantage
ⓐ No mechanical energy is converted to heat energy through friction.
ⓑ Some of the energy you put into rollerblading turns into heat energy through friction.

9. Machine
ⓐ a length of thread
ⓑ a pin

10. Mechanical advantage
ⓐ You can send a baseball farther by hitting it with a bat than you can by throwing it.
ⓑ If you pull a box straight up into a truck, you will not have to move it as far as you would if you pushed it up a ramp into the truck.

11. Power

ⓐ You can do six pull-ups in 30 seconds.

ⓑ You can do the flexed arm-hang longer than anyone else.

12. Resistance force

ⓐ You pull a weed with your hand.

ⓑ You dig out a weed with a shovel.

13. Actual mechanical advantage

ⓐ If the hills on the track were arranged correctly, a roller coaster could go on forever without an outside force.

ⓑ Unless there is an outside force, no hill on the roller coaster can be as high as the first one; as the coaster travels, some of its energy is converted to heat and sound.

14. Efficiency

ⓐ A ten-speed bike requires less work of the rider than a one-speed bike does.

ⓑ You and a friend have a bicycle race to determine who is faster. You win.

15. Friction

ⓐ The driver of a car hits the breaks and comes to a stop.

ⓑ The car will not start because it is out of gas.

16. Work input

ⓐ It is easier for you to push a box up a five meter ramp with the help of a friend than to do it alone.

ⓑ It takes 20 newtons of force to push a box up a five meter ramp.

17. Work output

ⓐ A pulley does 400 J of useful work.

ⓑ Friction in the pulley system converts some energy into heat energy.

Name: ______________________________ Date: ______________

Lab 4.3.1A Work, Force, and an Inclined Plane

QUESTION: How does an inclined plane affect work and force?

HYPOTHESIS: ____

____

EXPERIMENT:

You will need:
- 18 cm length of string
- pencil or pen
- shoe box
- brick or other heavy item
- Newton spring scale
- 1 m board
- chair
- protractor
- modeling clay
- meterstick

Steps:
1. Tie the string into a loop.
2. Use the point of a pen or pencil to poke a hole through the center of the end of the shoe box.
3. Thread the loop of string through the hole so part of the loop extends out of the hole and the rest remains inside.
4. Insert the pen or pencil through the part of the loop that is still inside the box. Pull the loop tight from the outside so the pencil holds the loop firmly in place.
5. Place the brick or other heavy item inside the shoe box.
6. Attach the spring scale to the loop of string that is on the outside of the box.
7. Set the board (inclined plane) against the seat of a chair so the angle where it meets the floor is approximately 30°. Check the angle using the protractor. Place a small amount of modeling clay on the floor to hold the board securely in place.
8. Measure the vertical distance from the floor to the point where the board rests on the chair seat. Record the distance.
9. Measure the distance along the board from the point where it meets the floor to the point where it rests on the chair seat. Record the distance.
10. Place the shoe box on the floor, and lift the box straight up using the spring scale. Record the force in newtons.
11. Place the shoe box at the base of the board, and pull the shoebox up the board using the spring scale. Record the force.
12. Repeat Steps 8–11, changing the angle to make the inclined plane less steep.
13. Repeat Steps 8–11, changing the angle to make the inclined plane steeper.

Vertical Distance (m)	Length of Incline (m)	Angle Between Floor and Ramp	Resistance Force (N)	Work (J) $F \times d$	Efficiency (%)

Lab 4.3.1A Work, Force, and an Inclined Plane

ANALYZE AND CONCLUDE:

1. Which is larger, the lifting force or the dragging force? ______________________

2. How does the dragging force change as the angle of the inclined plane is changed?

____

____

3. Which requires more work, lifting the object directly or dragging it up the inclined plane? ______________________

____

4. Does the efficiency of the inclined plane change as the angle changes? Why? ______

____

____

____

____

5. Why do people use inclined planes? ______________________

____

____

____

Name: ______________________________ Date: ______________

Lab 4.3.2A Classes of Levers

QUESTION: How does the relationship of the effort arm to the resistance arm affect mechanical advantage?

HYPOTHESIS: ______________________________

____

EXPERIMENT:

You will need:		
• cork with a nail in it • clamp	• retort stand • meterstick with a hole drilled at the 50 cm mark	• Newton spring scale • modeling clay • standard mass with a hook

Steps:

1. Secure the cork in the clamp so it does not wiggle.
2. Attach the clamp near the top of the retort stand.
3. Hang the meterstick on the nail through the hole in the meterstick.
4. Check that the meterstick is balanced. If not, add a small amount of modeling clay to the side that needs more weight so it balances.
5. Using the spring scale, measure and record the weight of the standard mass. ______________________________

First-Class Lever

6. Hold the lever level and hang the standard mass at a point 30 cm from the fulcrum. Attach a spring scale on the other side of the fulcrum, 30 cm from the fulcrum. Pull down on the spring scale. Measure the force necessary to balance the mass.
7. Leave the mass where it is, but move the spring scale to a point only 20 cm away from the fulcrum. Measure and record on the data table the force needed to balance the mass.
8. Leave the mass where it is, but move the spring scale to a point 40 cm away from the fulcrum. Measure and record the force needed to balance the mass.

Second-Class Lever

9. Ignore half of the meterstick. Consider that the fulcrum is at the end of the lever.
10. Hang a standard mass at a point 10 cm away from the fulcrum.
11. Attach a spring scale on the same side of the fulcrum at a point 40 cm away from the fulcrum.
12. Pull up on the spring scale. Record the force necessary to balance the mass.
13. Leave the mass where it is hanging. Move the spring scale so it is on the same side of the fulcrum as the mass at a point 30 cm away from the fulcrum.
14. Pull up on the spring scale. Record the force necessary to balance the mass.

Third-Class Lever

15. Ignore half of the meterstick, and consider the fulcrum to be the end of the lever.
16. Hang a standard mass 40 cm to one side of the fulcrum.
17. Attach a spring scale on the same side of the fulcrum, 30 cm away from it.
18. Pull up on the spring scale. Record the force necessary to balance the mass.

Lab 4.3.2A Classes of Levers

19. Leave the mass where it is hanging. Move the spring scale so it is on the same side of the fulcrum as the mass at a point 10 cm away from the fulcrum.
20. Pull up on the spring scale. Record the force necessary to balance the mass.

Class of Lever	Length of Resistance Arm (cm)	Length of Effort Arm (cm)	Effort or Input Force (N)	Mechanical Advantage (effort arm/resistance arm)

ANALYZE AND CONCLUDE:

1. What does a mechanical advantage of 1 mean? ______________________________

____

2. For the first-class lever, how did the effort change when the effort bar was shortened? ______________________________

____

3. To lift a heavy object, what would be the best way to build the first-class lever to get the greatest mechanical advantage? ______________________________

____

4. Using the lever from the lab, how could a first-class lever be set up to get a mechanical advantage of 3? ______________________________

____

____

5. How is a third-class lever different from a first- and a second-class lever? ________

____

____

____

Name: ______________________________ Date: ______________

Lab 4.3.2B Catapults

QUESTION: How does a catapult work and how can you increase distance?

HYPOTHESIS: ______________________________

____

EXPERIMENT:

You will need:	• tape	• meterstick
• paper target	• small projectile	

Steps:

1. Build the catapult according to directions on **BLM 4.3.2A Build a Catapult**.
2. Find the lever on the diagram and label the following: *rigid bar*, *fulcrum*, *resistance*, and *effort*.
3. Use arrows to label the direction of the effort or input force and the output force.
4. What class lever is the catapult? ______________________________
5. Tape the target to the wall.
6. Place the catapult about 1 m from the target.
7. Set the projectile on the catapult, pull down the arm, and fire.
8. Record distance and accuracy of the launch.
9. Make adjustments to improve the accuracy.
10. Launch again and record results.
11. Continue the process until the target is accurately hit three times in a row.

Trial	Distance Projectile Traveled	Accuracy (location of projectile relative to center of target)	Adjustments Made for Next Trial

Lab 4.3.2B Catapults

ANALYZE AND CONCLUDE:

1. What type of energy is used to move the catapult? ______________________

____

2. Calculate the mechanical advantage for the final design of the catapult. __________

____

3. How does that number show a mechanical advantage? ______________________

____

____

____

4. What energy does the projectile have when it is moving toward the target? ________

____

5. What force is acting on the projectile after it leaves the catapult? How does that affect its projectile motion? ______________________

____

____

6. How can the catapult be improved? ______________________

____

____

____

____

Name: ______________________________ Date: ______________

Lab 4.3.3A Pulling with Pulleys

QUESTION: How effective are pulleys at reducing the force needed to move objects?

HYPOTHESIS: ____

____

EXPERIMENT:

You will need:
- standard mass with hook
- Newton spring scale
- retort stand
- C clamp
- single fixed pulley
- meterstick
- 1 m piece of strong string
- single movable pulley
- double movable pulley

Steps:

1. Attach the standard mass to the spring scale. Lift the mass and record the force in newtons on the chart. This is the resistance force.
2. Secure the retort stand to the desk with the C clamp.
3. Suspend the single fixed pulley from the retort stand.
4. Hold a meterstick vertically on the desk so the distance the mass moves can be measured.
5. Tie one end of the string to the standard mass. Thread the string over the pulley, and attach the loose end to the spring scale.
6. Raise the resistance force a vertical distance of 25 cm. Record the force required to lift it. This is the effort force.
7. Measure the distance the string traveled to raise the load force 25 cm. Record this as the distance that the effort force moves.
8. Calculate the work put in to the machine and the work that the machine produces (output force).
9. Calculate the mechanical advantage and efficiency.
10. Change the setup to use a single movable pulley. Attach the load directly to the pulley. Attach one end of the string to the support rod; attach the other end to the newton spring scale. Repeat Steps 6–9
11. Change the setup to use both the single fixed pulley and the single movable pulley. Leave the resistance force attached directly to the hook on the bottom of the movable pulley, but attach the string to the hook on the top of the movable pulley. Repeat Steps 6–9.
12. Change the setup to use a single fixed pulley and a double movable pulley. Repeat Steps 6–9.

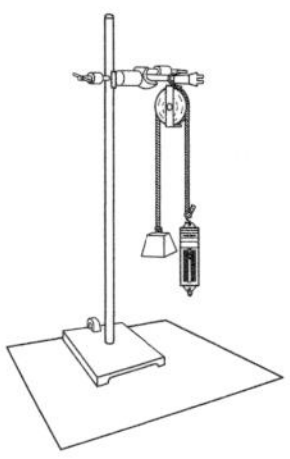

Single fixed pulley

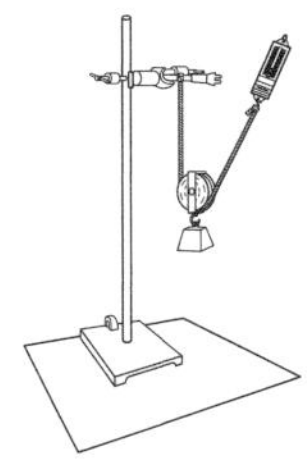

Single moveable pulley

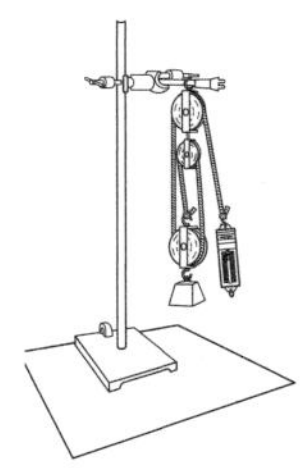

Single fixed and single moveable pulley

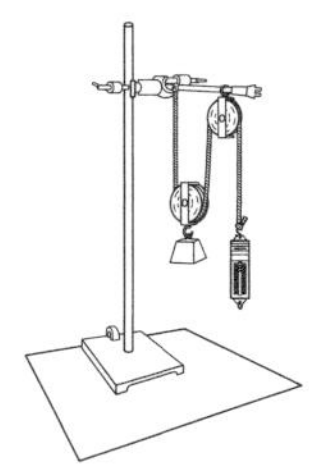

Double moveable pulley

Lab 4.3.3A Pulling with Pulleys

ANALYZE AND CONCLUDE:

1. What is the mechanical advantage (*MA*) of using a single fixed pulley? ______________

____

2. What is the advantage of using a single fixed pulley? ______________________________

____

3. Which system had the greatest *MA*? Why? ___________________________________

____

____

____

4. Which system is the most efficient? ____

5. Does the efficiency seem to be related to *MA*? Why? ____________________________

____

____

____

____

6. What relationship is there between the number of supporting ropes and the *MA* of a pulley system? ____

____

____

7. Where is the trade-off in a pulley system? ___________________________________

____

____

____

____

Name: ____________________ Date: __________

Lab 4.3.3A Pulling with Pulleys continued

Pulley Type	Resistance Force (N)	Distance Resistance (Load) Moves (m)	Effort Force (N)	Distance Effort Moves (m)	Work Input (W_1) in Joules	Work Ouput (W_2) in Joules	$MA = \frac{\text{load force}}{\text{effort force}}$	Efficiency = $\frac{W_2}{W_1} \times 100\%$
Single fixed								
Single movable								
Single fixed and movable								
Single fixed and double movable								

Name: ______________________ Date: __________

Lab 4.3.4A Gearing

QUESTION: How does the mechanical advantage of the gears affect the performance of a bicycle?

HYPOTHESIS: ______________________

____

EXPERIMENT:

You will need:
- masking tape
- marking pen
- board with 2 nails
- 4 spools, 3 with different diameters
- rubber band
- metric ruler

Steps:

1. Use the masking tape and the marking pen to label each nail on the board. Label one nail *Pedals* and the other nail *Wheel*.
2. Place one spool on each nail and connect them with a rubber band.
3. Make a reference mark on the edge of each spool. These marks will help measure the number of rotations as the spools turn.
4. Measure the diameter of each spool. Record the diameters.
5. The *Pedals* spool represents the chainring attached to the pedals on a bicycle. The *Wheel* spool represents the gear attached to the rear wheel. Using the reference marks, count the rotations of the *Wheel* spool as you rotate the *Pedals* spool five complete turns. Record the rotations.
6. Repeat Steps 3–5, replacing the *Wheel* spool with a different spool each time.
7. Calculate the ideal mechanical advantage (*IMA*) for each model, using the following formula, and record your answers on the table below.

$$IMA = \frac{5 \times \text{diameter of pedal}}{(\text{\# of wheel revolutions}) \times (\text{diameter of wheel})}$$

Pedals Spool Diameter (cm)	Revolutions of Pedals Spool	Wheel Spool Diameter (cm)	Revolutions of Wheel Spool	IMA
	5			
	5			
	5			

ANALYZE AND CONCLUDE:

1. Which combination of pedal and rear wheel gears gave the greatest mechanical advantage? Why? ______________________

____

Lab 4.3.4A Gearing

2. To increase speed, which combination would be best? Why? ______________________

____

____

____

____

3. To climb a steep hill, which combination would be the best? Why? ______________

____

____

____

____

Name: ______________________________ Date: ______________

Calculations

WS 4.3.1A

Solve the following problems. Convert units when necessary.

1. What is the resistance force when lifting a 24 kg box 3.6 m off the ground? How much energy is needed to lift it?

2. How much energy is needed to lift the same box 480 cm?

3. What is the ideal mechanical advantage (*IMA*) of a ramp that is 12 m long and reaches a height of 3.6 m?

4. If you use the same ramp to move the box and the effort force is 85 N, what is the work input?

5. Calculate the efficiency of the inclined plane that is 3.6 m off the ground.

6. What is the actual mechanical advantage (*AMA*) of the ramp? What does that mean?

7. How does the mechanical advantage change if the ramp is shortened to 9 m? What happens to the angle of the ramp?

8. What is needed to increase the mechanical advantage?

9. How is the work input affected by a longer ramp? Does the mechanical advantage change? Why?

Calculations **WS 4.3.1A**

10. Find the *IMA* of moving a couch with a mass of 3,550 g up an inclined plane with a height of 280 cm and an incline that is 1,042 cm long.

11. How much energy is needed to lift the couch 280 cm?

12. When using an inclined plane, how does the work input compare to the work output? Why?

13. If the effort force equals 20 N, what is the work input?

14. What is the efficiency of the inclined plane? What is the *AMA*?

15. How much energy would be needed to move 15 couches of the same mass up the same ramp as mentioned previously?

Name: ______________________ Date: __________

Life with Levers

WS 4.3.2A

Look around your house to find five items that use levers. List each device; diagram and label the parts of the lever; and identify it as a first-, second-, or third-class lever. Give a brief description of the item and how it is used.

1.

____

____

____

2.

____

____

____

3.

____

____

____

4.

5.

Name: ______________________ Date: __________

Mechanical Advantage

WS 4.3.4A

Solve the following problems:

1. Find the *IMA* of moving a table that has a mass of 2,120 g up an inclined plane that has a height of 240 cm and an incline that is 6 m long.

2. If the effort force in the previous problem equals 20 N, what is the work input?

3. If Josie uses one fixed pulley to lift a 100 kg container 5 m and she cannot move it, what can she do to triple her mechanical advantage?

4. Which ax-head will give Mark a greater mechanical advantage: an ax-head with a slope of 18 cm and a height of 6 cm, or an ax-head with a slope of 16 cm and a height of 4 cm? What is its *MA*?

5. What is the mechanical advantage of a screw with a pitch of 0.2 cm and a head with a diameter of 6 mm?

6. If the previous screw is replaced with a new screw that has a head of the same diameter but a pitch of 0.3 cm, what is the new *MA*? Is it better or worse?

7. Jan has a choice of two screwdrivers to use. One has a handle with a diameter of 4 cm and a shaft that is 16 cm. The other has a handle with a diameter of 5.5 cm and a shaft that is 10 cm. What is the *MA* of the better screwdriver if it is used on the screw in the previous question?

8. What is the mechanical advantage of using a wheel and axle that has a wheel diameter of 38 cm and an axle diameter of 0.12 m? How could the *MA* be improve without changing the axle?

Name: ____________________ Date: __________

Small Wonders and Big Deals

WS 4.3.5A

Write a sentence or two about how these complex machines, or the electronic versions that came after them, affect the way that people live today.

1. Sewing machine (invented in 1851): ____________________

____

____

2. Clock with a minute hand (invented in 1577): ____________________

____

____

3. Printing press (invented in 1450): ____________________

____

____

4. Reaper (invented in 1831): ____________________

____

____

5. Vending machines (invented in the early 1800s): ____________________

____

____

6. Elevator brake (invented in 1854): ____________________

____

____

7. Flush toilet (invented in 1738): ____________________

____

____

8. Can opener (invented in 1858): ____________________

____

____

9. Calculator (invented in 1642): ____________________

____

____

Name: ______________________ Date: __________

Business Machines

WS 4.3.5B

Answer the following questions according to the observations made during a trip to a local business:

1. Give an example of a machine that includes a lever. Name the part of the machine functioning as a lever. ______________________

____

____

2. Give an example of a machine that includes an inclined plane. Name the part functioning as an inclined plane. ______________________

____

____

3. What is an example of a machine that includes a wheel and axle? What is the name of the part? ______________________

____

____

4. Give an example of a machine that includes a pulley. Which part functions as a pulley? ______________________

____

____

5. What is an example of a machine that uses a wedge? Name the part that functions as a wedge. ______________________

____

____

6. Sketch or name any devices that use hydraulic or pneumatic forces. __________

____

____

____

Name: ______________________________ Date: ______________

Lab 5.1.2A Calorimetry

QUESTION: Which substance has the greatest specific heat?

HYPOTHESIS: ______________________________

____

EXPERIMENT:

You will need:

- post-1982 US pennies, 100 g
- nickels, 100 g
- aluminum nails, 100 g
- mass scale
- masking tape
- 8 large foam cups
- marking pen
- 4 thermometers
- large beaker
- hot plate
- heat-resistant gloves
- tongs
- towel

Steps:

1. Measure 100 g each of pennies, nickels, and aluminum nails.
2. Stack the pennies and bind them together with masking tape.
3. Stack the nickels and bind them together with masking tape.
4. Bind the nails together with masking tape.
5. Stack the foam cups in groups of two for a total of four sets.
6. Label the cups: *Pennies*, *Nickels*, *Nails*, and *Water*.
7. Pour 300 g (300 mL) of room-temperature water into each set of cups.
8. Place a thermometer in each cup and record the temperatures.
9. Heat a large beaker of water to boiling. Carefully place the pennies, nickels, and nails in the boiling water. Leave them in the water for about 8 minutes so they heat to the boiling point as well.
10. Wearing heat resistant gloves, remove the pennies with tongs. Quickly and carefully pat the pennies dry, and lower them into the cup labeled *Pennies*. Observe the rise in temperature on the thermometer. Record the highest temperature.
11. Using the same procedure, put the heated nickels in the *Nickels* cup and the nails in the *Nails* cup. Record the highest temperature that the thermometer in each cup reaches.
12. Carefully pour 100 mL of boiling water into the cup labeled *Water*. Record the highest temperature the thermometer measures.

Calorimetry

Heated Substance	Initial Water Temperature	Final Water Temperature	Change in Temperature
Pennies			
Nickels			
Nails			
Water			

Lab 5.1.2A Calorimetry

ANALYZE AND CONCLUDE:

1. Compare the masses of the substances. ______

2. What happened to the heat that was absorbed by each substance? ______

3. Do the results confirm the law of energy conservation? Why? ______

4. List the tested substances in order according to their specific heat. Begin with the substance with the highest specific heat. ______

Name: ______________________________ Date: ______________

Lab 5.1.3A Thermal Insulators

QUESTION: Which substance is the best insulator?

HYPOTHESIS: ____

____

EXPERIMENT:

You will need:
- pan
- hot plate
- large foam cup with lid
- 2 beakers
- 3 thermometers
- nail
- square of heavy wool fabric
- 3 rubber bands
- scissors
- square of aluminum foil
- timer
- access to a refrigerator

Steps:

1. Pour a liter of water into the pan. Place the pan on the hot plate and bring the water to a boil.
2. Pour 200 mL of the hot water into the foam cup and each of the two beakers.
3. Measure the temperature of the water in each container and record the data.
4. With a nail, punch a hole in the cup lid. Put the lid on the cup and place the thermometer through the hole.
5. Put the wool fabric over the top of the first beaker. Secure the fabric around the beaker with rubber bands. Place a thermometer in the beaker through a hole cut into the fabric.
6. Put the aluminum foil over the top of the second beaker. Press the foil over the edges of the beaker. Place a thermometer in the beaker through a hole punched in the foil.
7. Place the three containers in a refrigerator. Record the temperature of each every 2 minutes for 14 minutes.

	Foam Cup	Beaker with Wool	Beaker with Foil
Starting Temp.			
2 minutes			
4 minutes			
6 minutes			
8 minutes			
10 minutes			
12 minutes			
14 minutes			

Lab 5.1.3A Thermal Insulators

ANALYZE AND CONCLUDE:

1. At the bottom of this page, make a line graph that shows the change in temperature over time for all three containers. Which variable will be on the x-axis? Which variable will be on the y-axis? How many lines are needed and what labels will they have? ______________________________

____

____

____

2. Which was the best insulator? Why? ______________________________

____

____

3. Which was the worst insulator? Why? ______________________________

____

____

4. What can good insulators be used for? ______________________________

____

____

5. What can poor insulators be used for? ______________________________

____

____

____

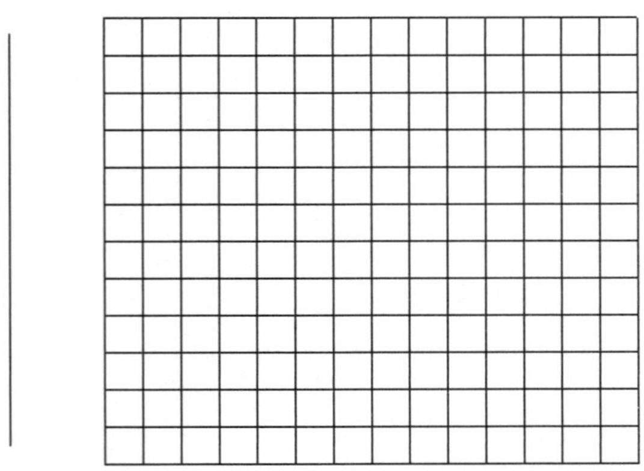

0

____

Name: ____________________ Date: ____________

Lab 5.1.3B Radiation

QUESTION: Which side of the can will absorb more heat?

HYPOTHESIS: ____________________

____

EXPERIMENT:

You will need:		
• hammer	• 10 toothpicks	• matches
• small nail	• paintbrush	• lamp with incandescent bulb
• large tuna or chicken can	• black paint	
	• candle	

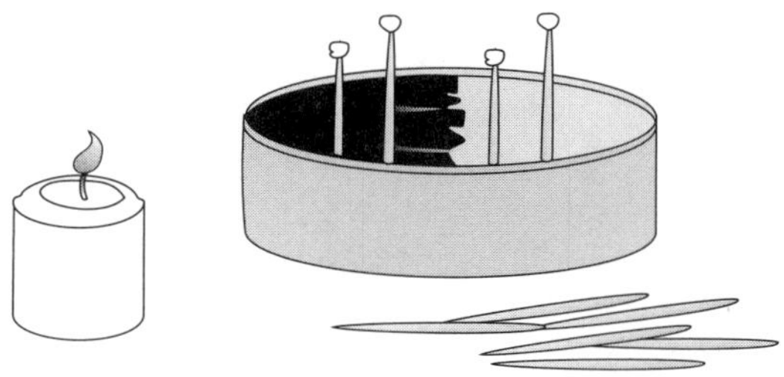

Steps:

1. Using the hammer and nail, punch 10 holes in the bottom of the can. Verify that a toothpick will fit firmly into each hole.
2. Being careful about any sharp places on the can, paint half of the bottom and half of the inside of the can black.
3. Light the candle.
4. Drip candle wax on one end of each toothpick. Allow the wax to harden.
5. Once the paint in the can is dry, place the wax-tipped toothpicks into the holes in the bottom of the can, with the wax tips pointing up.
6. Turn on the lamp. Invert the can, and hold the bottom of the can over the light bulb.
7. Continue until the wax starts melting on at least two toothpicks.

ANALYZE AND CONCLUDE:

1. Which side of the can had wax melt first? ____________________

____

2. Use the term *radiation* to explain your observations. ____________________

____

____

____

____

Name: ______________________________ Date: ______________

Lab 5.1.4A Heat and Work

QUESTION: How will chemical energy be converted into mechanical energy?

HYPOTHESIS: ______________________________

____

EXPERIMENT:

You will need:	• quick-drying caulk or rubber cement	• unsen burner
• hammer	• 500 cm length of thin string	• ring stand
• large nail	• duct tape or electrical tape	• coffee can with lid
• metric ruler	• two 5–6 cm lengths of aquarium hose	

Steps:

1. Use the hammer and nail to punch two holes on opposite sides of the can.
2. Insert a piece of hose into each hole so half of the hose is in the can and the other half is sticking out of the side.
3. Use the caulk or rubber cement to seal the openings around each tube. Bend each tube in a clockwise direction.
4. Pound a small hole in the center of the lid. Tie a knot at the end of the string and draw the string through the lid, pulling the knot tightly against the lid. Seal the opening with caulk or rubber cement.
5. Pour water into the can to a depth of 2–3 cm. Put the lid on the can and tape around the edge.
6. Tie the can to the ring stand, and set the can just above the Bunsen burner.
7. Heat the water in the can.

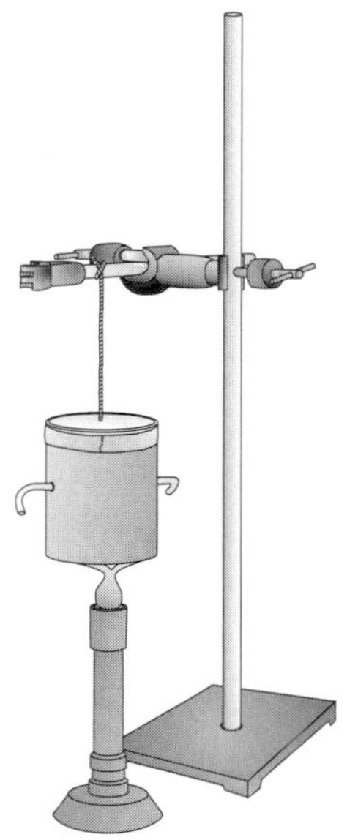

ANALYZE AND CONCLUDE:

1. What happens when the water begins to heat up? ______________________________

2. Why does this happen? ______________________________

____

____

3. Explain the transfer of energy, beginning with the natural gas in the Bunsen burner. ______________________________

____

____

____

Lab 5.1.4A Heat and Work

4. How can the steam be harnessed to do work? ______

______

______

______

______

Name: ______________________________ Date: ______________

Heat and Temperature **WS 5.1.1A**

Answer the following questions:

1. What things are made of particles? What things are not made of particles? ________

2. What is the kinetic energy of a nonmoving 7 kg bowling ball that is in the middle of a bowling alley? ________

3. What is the kinetic energy of a 7 kg object sliding 2 m/sec across the floor? What if it were sliding at 4 m/sec across the floor? ________

4. What is heat? ________

5. What is the difference between heat and temperature? ________

6. Explain why metal pans get warm over a campfire. ________

7. What is the evidence of heat in a system? ________

8. What will eventually happen to the system? ________

9. Will the system have any kinetic energy? How could it be tested for the energy?

Name: ______________________________ Date: ______________

Lab 5.2.1A Isotopic Pennies

QUESTION: How can the mass of specific isotopes be determined from a sample of multiple isotopes?

HYPOTHESIS: ______________________________

____

EXPERIMENT:

You will need: • mass scale • pre-1982 US penny	• post-1982 US penny • empty film canister	• 10 pennies of various dates

Steps:

1. Using the mass scale, determine the individual masses of the pre-1982 penny, the post-1982 penny, and the empty film canister. ______________________________

   ____
2. Multiply the mass of the older penny by 10 to determine what the mass of the hidden pennies would be if they were all pre-1982 pennies. ______________
3. Multiply the mass of the newer penny by 10 to determine what the mass of the hidden pennies would be if they were all post-1982 pennies. ______________
4. Put the 10 pennies of various dates inside the film canister. Use the scale to determine the mass of the container of pennies.
5. Subtract the mass of the empty canister. This is the "isotopic" mass of the two different "isotopes" of pennies. ______________________________

ANALYZE AND CONCLUDE:

1. What would be the isotopic mass if all the pennies were pre-1982? ______________

2. What would be the isotopic mass if all the pennies were post-1982? ______________

3. If the sample contained five pre-1982 and five post-1982 pennies, what would be the isotopic mass? ______________________________

____

4. Did the canister have more pre-1982 pennies or post-1982 pennies? Why? ________

____

____

____

Lab 5.2.1A Isotopic Pennies

5. Use the following formula to determine how many pennies of each kind were in the sample.
n = # of older pennies
$10 - n$ = # of newer pennies
total mass = (n × mass of older penny) + ([$10 - n$] × mass of newer penny)

6. What determines whether two isotopes are the same element? ____________________

____

7. Do all isotopes of a given element have the same atomic mass? ____________________

____

Name: ______________________ Date: __________

Lab 5.2.4A Half-Life Probability

QUESTION: How much isotope remains after each half-life?

HYPOTHESIS: ______________________

____

EXPERIMENT:

You will need: • 100 candy coated chocolates with a letter stamped on one side	• large cup • sheet of paper • colored pencils	• graph paper

Steps:

1. Place all candies so their printed side is down on the paper. These represent atoms of the radioactive isotope before any decay starts.
2. Pour all the candies into the large cup and gently shake.
3. Carefully pour the candies back on to the paper. Count the number that land facedown. These are the "survivors," or the isotopes that have not decayed yet. Record the number. Record the number of survivors for the other groups as well. This represents the amount of original isotope that still exists after one half-life.

Number of "Surviving" Isotopes

Half-Life	Class Total	Group 1	Group 2	Group 3	Group 4	Group 5	Group 6	Group 7	Group 8	Class Average
1										
2										
3										
4										
5										
6										
7										
8										

4. Place the decayed isotopes back in the cup and shake again. Pour them on the paper. Count the number of non-decayed isotopes and record. Record all groups' results as well. This is the second half-life.
5. Return the decayed isotopes to the cup and continue the process five more times or until there is no more isotope to decay.
6. Calculate the class totals and averages for each half-life and record.
7. Create a graph showing the amount of surviving isotope for each half-life.

Lab 5.2.4A Half-Life Probability

8. Using a colored pencil, plot the results for your group and connect with a line.
9. Plot the class average results with a different color and a thicker line.
10. Using a third color, plot the expected quantities of isotope remaining for each half-life. For example, the starting isotope quantity for zero half-life is 100. For the first half-life, the expected point is 100/2, or 50. The second half-life would be 50/2, or 25.

ANALYZE AND CONCLUDE:

1. Why did each group not get the same results? ______________________________

____

2. Which follows the expected results better, your group's results or the class average? Why? ____

____

3. Is a half-life an absolute value or an estimated value? Why? ______________

____

____

Name: ______________________________ Date: ______________

Isotopes

WS 5.2.1A

1. Graph the nonradioactive isotopes, using the atomic number (the first number) for the x-axis and the mass number (the second number) for the y-axis. Connect these points with a blue pencil.

2. Graph the radioactive isotopes, using the atomic number (number of protons in the atom) for the x-axis, and the mass number (number of protons and neutrons in the atom) for the y-axis. Connect these points with a red pencil.

Nonradioactive and Radioactive Isotopes

Elements	Hydrogen H	Carbon C	Sodium Na	Potassium K	Iron Fe	Cobalt Co	Rubidium Rb	Iodine I
Nonradioactive Isotopes (mass number, atomic number)	(1, 1)	(6, 12)	(11, 23)	(19, 39)	(26, 56)	(27, 59)	(37, 85)	(53, 127)
Radioactive Isotopes (mass number, atomic number)	(1, 3)	(6, 14)	(11, 24)	(19, 40)	(26, 59)	(27, 60)	(37, 87)	(53, 131)

Isotope Comparison

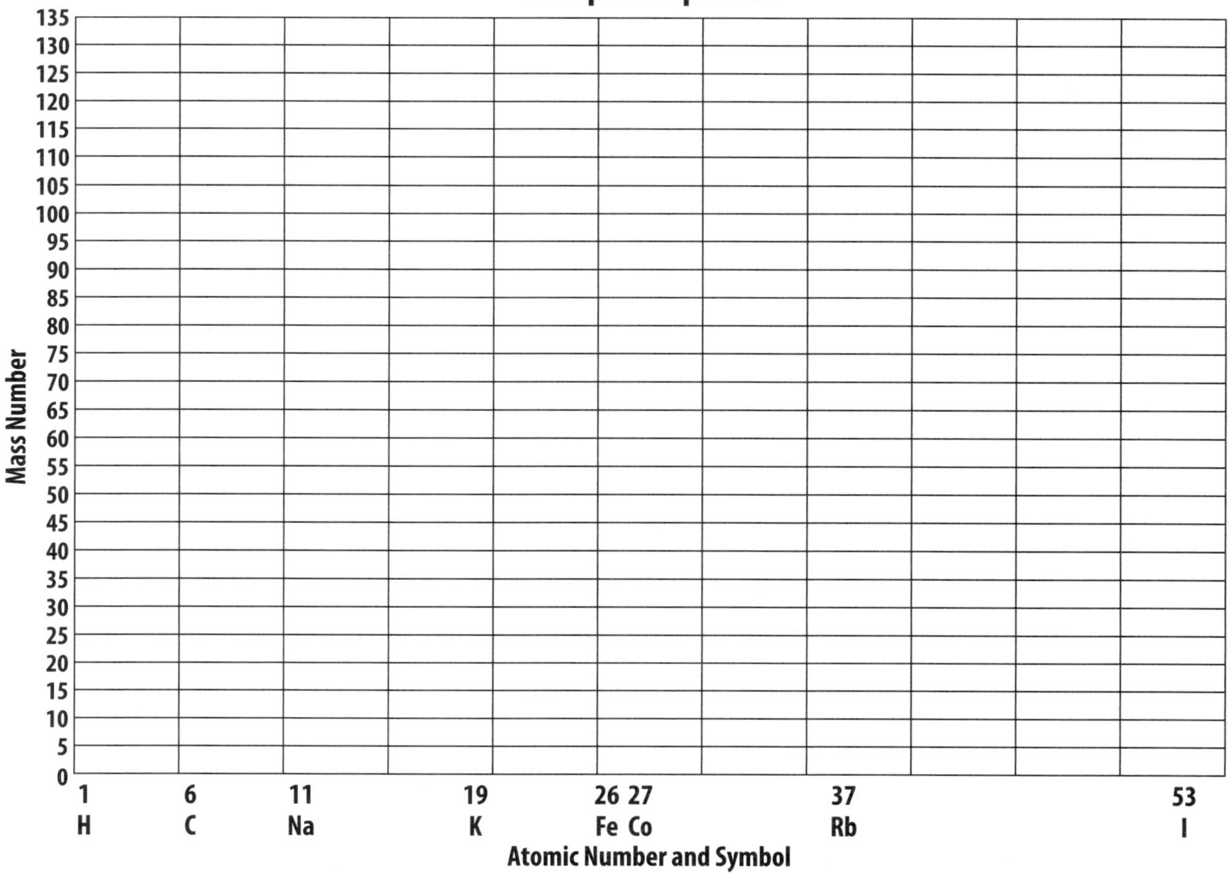

Isotopes

WS 5.2.1A

3. Do the two lines cross over each other? Why? ______________________

____

____

____

4. Given the data, what conclusion can be made about radioactive isotopes? __________

____

____

Name: ______________________________ Date: ______________

Transmutation

WS 5.2.2A

Uranium-238 (U-238) transmutates 15 times before turning into the stable isotope lead-206 (Pb-206). These 15 steps are known as *the decay series* of this isotope.

1. Use the data below to graph the transmutation of U-238. Plot the atomic numbers on the x-axis and the mass numbers on the y-axis.

2. Next to each plotted point, write the isotope symbol of the newly transmuted isotope. For example, next to the first step in the radioactive decay of U-238, identify the new isotope as *234/90Th*.

Decay Series of Uranium-238

Atomic Number	Mass Number
92	238
90	234
91	234
92	234
90	230
88	226
86	222
84	218
82	214
83	214
84	214
82	210
83	210
84	210
82	206

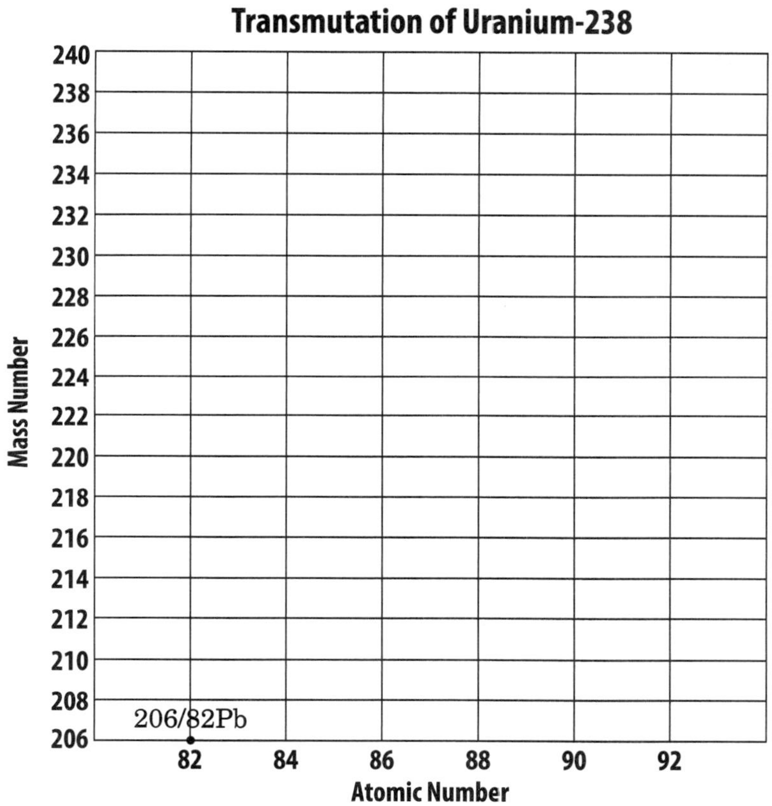

3. If a substance went through alpha decay, what happened to the mass number and the atomic number? ______________________________

____

4. If a substance went through beta decay, what happened to the mass number and the atomic number? ______________________________

____

5. Which number always changes in alpha and beta decay? Why? ______________

____

____

Name: ______________________ Date: ____________

World Map

WS 5.2.3A

Find the locations of uranium mines across the world and plot them on the map.

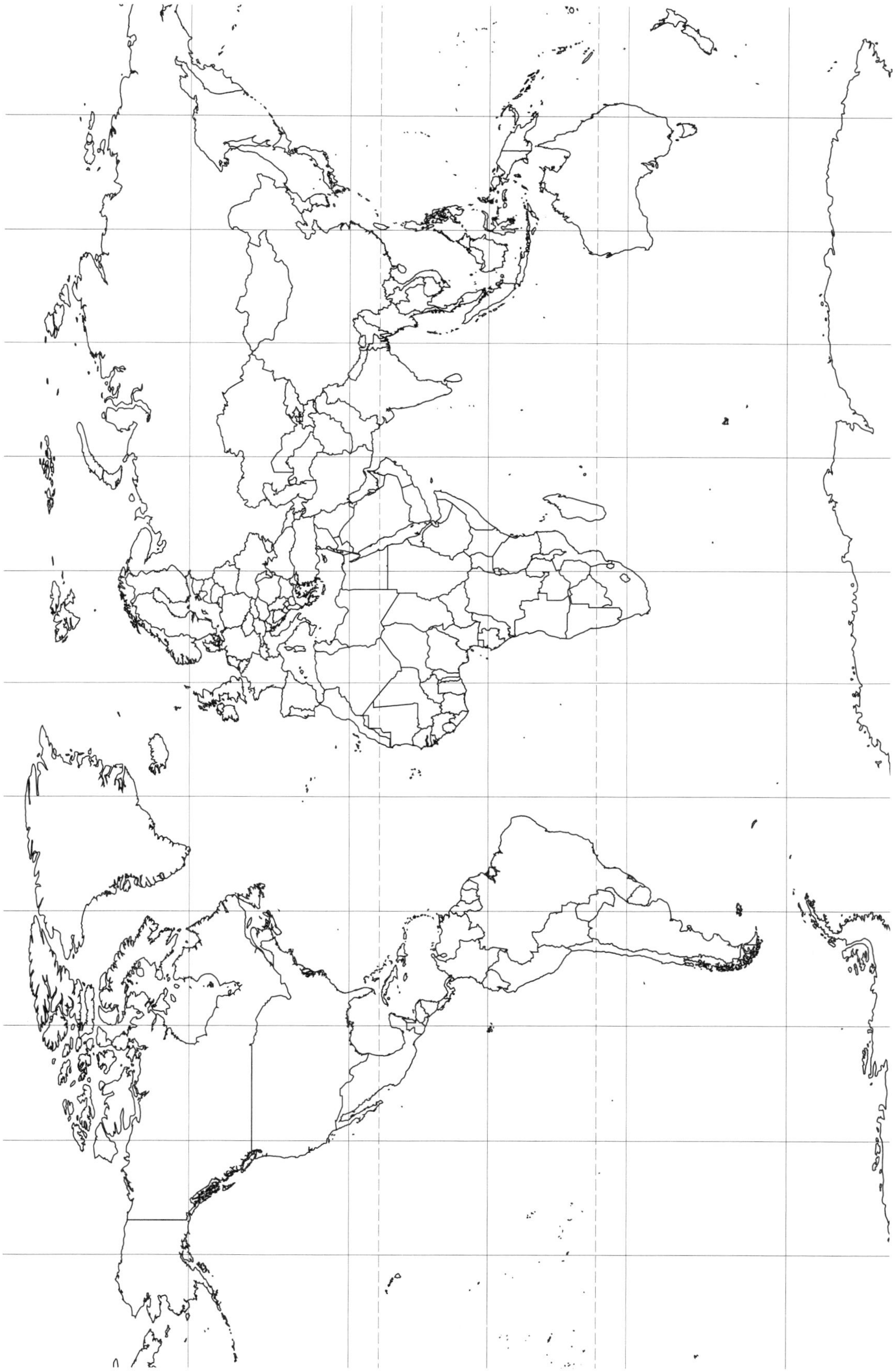

1. Which country produces the greatest amount of uranium? ______________________

2. How much uranium did it produce in 2016? ______________________

Name: ______________________ Date: ____________

Nuclear Power Plant

WS 5.2.3B

Answer, and write additional questions to ask and answer.

1. How much energy does this power plant produce? ______________________

2. What safety precautions are in place? ______________________

3. How much does it cost to run the power plant for one year? ______________________

4. How much does the nuclear power cost compared with other power sources in the region? ______________________

5. What materials are used to generate the power? ______________________

6. To what extent did you need to educate the public when the nuclear power plant was first constructed? How much do you communicate with the public now? ______________________

7. ______________________

8. ______________________

9. ______________________

Name: ______________________________ Date: ______________

Lab 5.3.3A Energy from the Sun

QUESTION: Which container will gain the most heat energy?

HYPOTHESIS: ____

____

EXPERIMENT:

You will need:	• 2 sheets white construction paper	• scissors
• tape	• 2 metal or plastic containers of similar size with plastic lids	• 2 thermometers
• 2 sheets black construction paper		

Steps:

1. Tape two layers of black paper around the first container and two layers of white paper around the second container.
2. With the scissors, punch a hole in each lid so a thermometer will fit through.
3. Fill each container to the same level with water. Cover them with the lids.
4. Insert a thermometer through each lid. The bulb should be just below the water's surface.
5. Place the containers in direct sunlight. Record the initial temperature.
6. Record the temperature of the water in each container every five minutes for 30 minutes.

Container	Initial Temperature	5 minutes	10 minutes	15 minutes	20 minutes	25 minutes	30 minutes
Black paper							
White paper							

7. Graph the data, plotting time on the x-axis and temperature on the y-axis.
8. Connect the temperature points of each container with a different colored pencil.

ANALYZE AND CONCLUDE:

1. In which container did the temperature rise most quickly? Why? ____________________

____

____

____

2. How could the results of this activity be used in a practical way? ____________________

____

____

____

Name: ____________________ Date: ____________

Tally Your Resources WS 5.3.1A

Look at 10 electrical appliances in your home, and list the power rating of each. The power rating is found on the back of the appliance.

Appliance	Power Rating

1. Which appliance has the highest rating? What is it? ____________________

2. Which appliance has the lowest rating? What is it? ____________________

3. Which appliance is used the most? ____________________

4. What can be done to lower the energy usage at home? ____________________

Name: ______________________________ Date: ____________

The Gift of Fossil Fuels

WS 5.3.2A

Imagine life without fossil fuels. List 20 things throughout the day that would be different without fossil fuels. These may include changes in travel, items that were manufactured using fossil fuels, and items that were manufactured long distances away and brought in through the power of fossil fuels.

1.	11.
2.	12.
3.	13.
4.	14.
5.	15.
6.	16.
7.	17.
8.	18.
9.	19.
10.	20.

1. What benefits might come from reducing or eliminating the use of fossil fuels?

____

____

____

____

2. What benefits might be lost from eliminating the use of fossil fuels? ____________

____

____

____

____

Name: ________________________________ Date: ______________

Classify Resources

WS 5.3.3A

Each group of terms contains a word that is not related to the others. Fill in the circle of the unrelated term, and explain why it does not belong.

1. ⓐ coal
ⓑ petroleum
ⓒ natural gas
ⓓ hydroelectric power

____

____

2. ⓐ anthracite
ⓑ biomass
ⓒ lignite
ⓓ bituminous coal

____

____

3. ⓐ hydroelectric power
ⓑ hydrocarbons
ⓒ solar energy
ⓓ geothermal energy

____

____

4. ⓐ geothermal energy
ⓑ biomass
ⓒ coal
ⓓ natural gas

____

____

Name: ______________________________ Date: ______________

Lab 6.1.1A Wave Mediums

QUESTION: How does the type of medium affect the wave speed?

HYPOTHESIS: ______________________________

____

EXPERIMENT:

You will need:		
• 1 m of thin rope	• timer	• metal extendable spring toy
• meterstick	• 1 m of thick rope	
	• plastic extendable spring toy	

Steps:

1. Extend a thin rope on a flat surface between you and your partner. Do not pull the rope tight.
2. Using a meterstick, measure the distance between you and your partner.
   - Distance: ________ m
3. Practice generating a few transverse waves with the rope. Make sure the waves travel the full distance of the rope.
4. Using a timer, record the time it takes for one wave to travel the length of the rope.
   - Time: ________ seconds
5. Repeat Steps 1–4 using a thick rope, a plastic extendable spring toy, and a metal extendable spring toy. Record your data in the table below.

Material	Distance (m)	Time (sec)	Speed (m/sec)
Thin rope			
Thick rope			
Plastic extendable spring toy			
Metal extendable spring toy			

ANALYZE AND CONCLUDE:

1. Calculate the speed of each wave by dividing the distance by time. ($v = d/t$) Record the calculations in the table.

2. Did the waves travel at the same speed through the different mediums? Explain.

____

____

____

____

____

Lab 6.1.1A Wave Mediums

3. Compare the shape of the waves created by the different mediums. ____________________

____

____

____

____

4. How did each wave gain energy? ____________________________________

____

____

____

Name: ______________________________ Date: ______________

Lab 6.1.3A Wave Interactions

QUESTION: How do waves respond to different barriers?

HYPOTHESIS: ______________________________

____

EXPERIMENT:

You will need:	• wooden dowel	• rectangular piece of glass
• large shallow pan	• 2 wooden blocks	• metric ruler
• wave fence	• protractor	

Steps:

1. Line the inside of the pan with the wave fence to decrease the reflection of waves off the sides of the pan.
2. Place a wooden dowel on the surface of the water, parallel and close to one of the short ends of the pan.
3. Allow the water to become still.
4. Without touching the water, move the dowel about 2 cm toward the other end of the pan. Record your observations in the table on the next page.
5. Place a wooden block at the end of the pan, opposite the dowel.
6. Allow the water to become still.
7. Without touching the water, move the dowel about 2 cm toward the wooden block. Record your observations in the table on the next page.
8. Adjust the placement of the wooden block to form a 45° angle with the incoming waves. Allow the water to become still.
9. Without touching the water, move the dowel about 2 cm toward the wooden block.
10. Observe the angle of incidence and the angle of reflection.
11. Determine where the normal is located. Using a protractor, estimate the size of the two angles. In the space provided here, draw a diagram of the reflected waves that are created. Label the normal, angle of incidence, and angle of reflection.
12. Remove the wooden block and replace it with an upright, rectangular piece of glass. Allow the water to become still.
13. Without touching the water, move the dowel about 2 cm toward the piece of glass. Record your observations in the table on the next page.

Lab 6.1.3A Wave Interactions

14. Remove the piece of glass, and replace it with two wooden blocks that are parallel to the incoming waves. Ensure there is about a 3 cm space left open between the two blocks.
15. Without touching the water, move the dowel about 2 cm toward the two wooden blocks. Record your observations in the table below.
16. Move the blocks closer and farther apart from one another to examine which space size creates the largest diffraction.

Step	Observation
Step 4: No barrier	
Step 7: One parallel wooden block	
Step 13: Glass	
Step 15: Two wooden blocks	

ANALYZE AND CONCLUDE:

1. How do the angle of incidence and the angle of refraction relate? ______

2. What type of wave behavior was demonstrated with the piece of glass? Why?

______

3. How did the distance between the two blocks affect the waves that were produced?

______

4. What would happen if waves were created at both ends of the pan? How would they interact? ______

Name: ______________________________ Date: ______________

Wave Speed

WS 6.1.2A

1. Label the following parts of a transverse wave on the diagram below: *crest*, *trough*, *wavelength*, *amplitude*, and *wave height*.

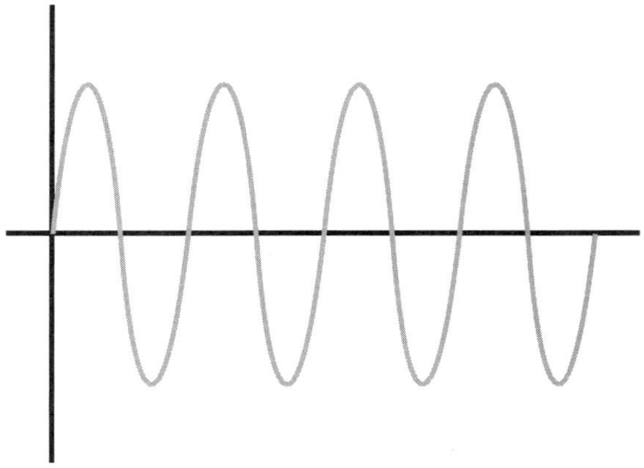

2. Fill in the missing information in the table below.

Medium	Frequency (Hz)	Wavelength (m)	Wave Speed (m/s)
Aluminum coils	1.2		4.7
Aluminum coils	0.8		4.7
Iron coils	5.3	0.6	
Iron coils		1.2	3.2

Use the table above to answer Questions 3–4.

3. How is the wave speed affected if the frequency in a medium decreases?

____

4. What happens to the wavelength of a medium when the frequency increases?

____

5. If two coils have wavelengths in a ratio of 1 to 3, what is the ratio of frequencies for these same two coils? ______________________________

6. Vivian and Moira create transverse waves with an extendable spring toy. They are standing 10 m apart. The wave height of the toy is 50 cm, the frequency is 4.5 Hz, and the horizontal distance from one crest to a nearby trough is 73 cm. What are the wavelength, period, amplitude, and speed (m/s) of the wave?

Wave Speed

WS 6.1.2A

7. Weather forecasters are anticipating a hurricane because oceanographers have been observing an increase in ocean wave swells. They estimate the wave height is 7.2 m and the distance between adjacent crests is 10.1 m. Waves are crashing onto shore every 9.7 seconds. Calculate the frequency and speed of these waves.

8. Phuong and Bic are floating next to each other in separate kayaks. They are 5 m apart. When a wave crest is under Phuong's kayak, a wave trough is under Bic's kayak. They float up and down every 4 seconds. Calculate the speed of the waves.

Name: ______________________________ Date: ____________

Wave Knowledge

WS 6.1.3A

Read each statement below. If the statement is true, circle the letter under the true column. If it is false, circle the letter under the false column. Use the letters and the question numbers to crack the code and answer the question.

	Statement	True	False
1.	A wave will travel more quickly through water than through brick because water returns more quickly to its original shape.	C	E
2.	A wave in deep water is a longitudinal wave.	M	H
3.	The tightly packed springs in a extendable spring toy make up a compression.	A	C
4.	A transverse wave travels side-to-side like a snake.	R	N
5.	A wave transfers matter through space.	W	G
6.	The distance from a medium when it is at rest to the highest point is its amplitude.	T	W
7.	The highest point of a wave is called its crest.	E	A
8.	The number of waves that pass a particular point in a unit of time is its frequency.	C	B
9.	The hertz is the unit used to measure wave period.	N	L
10.	On a transverse wave, the trough is the lowest point.	Y	S
11.	On a transverse wave, the wave height is the highest point.	S	Y
12.	The longer the wavelength, the shorter the wave period.	G	Y
13.	The energy from a light wave that undergoes absorption may be converted into heat.	R	I
14.	The angle of incidence is greater than the angle of reflection.	I	E
15.	Two waves that combine to form a wave with a larger amplitude than the original waves experience constructive interference.	R	S
16.	Two waves that combine to form a wave with a smaller amplitude experience destructive interference.	L	A
17.	Diffraction is the bouncing of a wave off a barrier.	R	N
18.	A refracted wave is completely absorbed; a reflected wave bends because of a change in speed.	R	A

What role do all waves play in creation?

6	2	14	10

3	16	9

8	18	13	4	12

1	17	7	15	5	11

Name: ______________________________ Date: ______________

Lab 6.2.1A Sound in a Vacuum

QUESTION: Can sound travel in a vacuum? Explain.

HYPOTHESIS: ______________________________

____

EXPERIMENT:

You will need:	• small bell	• 100 mL graduated cylinder
• tack	• short piece of string	• hot plate
• stopper	• 250 mL Erlenmeyer flask	• heat-resistant gloves

Steps:

1. Push a tack into the base of a stopper.
2. Attach the bell to the tack with a short piece of string.
3. Insert the bell into the flask, and seal the flask with the stopper. Ensure that the bell hangs in the middle of the flask and does not come in contact with the bottom of the flask.
4. Predict if the sound of the bell will be audible when you jostle the flask. Record your hypothesis and explain your reasoning. ______________________________

   ____
5. Gently shake the flask. Record your observations. ______________________________

   ____
6. Remove the stopper and the bell.
7. Using a graduated cylinder, pour 20 mL of water into the flask.
8. Place the flask on a hot plate and gently heat the flask until the water boils. Boil the water for two minutes.
9. Turn off the hot plate.
10. While wearing heat-resistant gloves, carefully insert the bell into the flask and seal the flask with the stopper.
11. Allow the flask to cool.
12. Gently shake the flask. Record your observations. ______________________________

    ____

ANALYZE AND CONCLUDE:

1. Compare the sound of the bell in Steps 5 and 12. ______________________________

____

2. What accounted for the difference in sound? ______________________________

____

____

Lab 6.2.1A Sound in a Vacuum

3. What mediums did the sound waves travel through in this experiment? ______________

____

4. Why do sound waves require a medium to travel? ______________________________

____

____

Name: ______________________ Date: __________

Lab 6.2.2A The Doppler Effect

QUESTION: How does sound change when the source or listener is moving?

HYPOTHESIS: ______________________________

____

____

____

EXPERIMENT:

You will need:		
• scissors	• one-hole punch	• piezoelectric buzzer
• plastic soda bottle	• 3 m nylon cord	• duct tape or electrical tape
	• 9 V battery	

Steps:

1. With a pair of scissors, carefully remove the upper third portion of a soda bottle. Do not dispose of the cut-off portion.
2. Using a one-hole punch, punch two holes about 2 cm below the edge of the bottle on opposite sides of the bottle.
3. Tie the ends of the cord through the punched-out openings.
4. Position the 9 V battery and the piezoelectric buzzer side-by-side inside the soda bottle. Using duct tape or electrical tape, secure the battery and buzzer in place. Attach the buzzer wires to the battery terminals and secure them with a piece of tape.

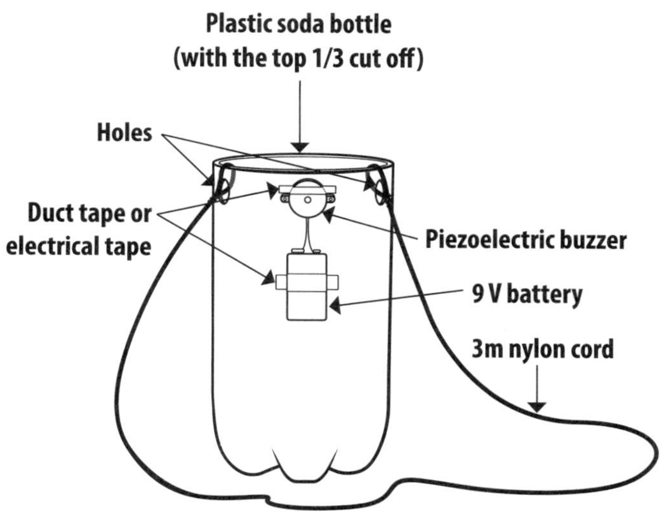

5. Stand a safe distance away from your laboratory partner and other laboratory groups.
6. Carefully swing the bottle above your head in a circular motion.
7. Have your laboratory partner note the sound differences.

ANALYZE AND CONCLUDE:

1. Is the pitch higher when the buzzer is approaching or moving away from the listener? ______________________________

____

____

Lab 6.2.2A The Doppler Effect

2. Explain this observation. ______________________________

____

____

3. What is this called? ______________________________

4. Why is the Doppler effect usually noticeable only with human-made sounds? What are some exceptions? ______________________________

____

____

____

____

Name: ______________________________ Date: ______________

Lab 6.2.3A Musical Tones

QUESTION: How does resonance affect sound?

HYPOTHESIS: ______________________________

____

EXPERIMENT:

You will need:	• test tube rack	• graduated cylinder
• 6 test tubes		

Steps:

1. Place six test tubes in a test tube rack.
2. Do not fill the first test tube.
3. Using a graduated cylinder, add 5 mL of water to the second test tube.
4. Add 10 mL of water to the third test tube, 15 mL of water to the fourth test tube, 20 mL of water to the fifth test tube, and 25 mL of water to the sixth test tube.
5. Predict the pitch of the sound that will be made when air is blown across the top of each test tube. Record your hypothesis. ______________________________
6. Blow across the top of each test tube.
7. Record your observations. ______________________________

ANALYZE AND CONCLUDE:

1. Which test tube created the highest pitch? ______________________________

2. Which test tube created the lowest pitch? ______________________________

3. Why did the pitch change with varying amounts of water? ______________________________

4. Which air column length produced the highest sound frequencies? ______________________________

Lab 6.2.3A Musical Tones

5. How does resonance affect the sound produced in each test tube? ______________

____

____

____

Name: ______________________________ Date: ______________

Lab 6.3.2A The Path of Light

QUESTION: How does light travel?

HYPOTHESIS: ______________________________

____

EXPERIMENT:

You will need:		
• metric ruler	• 5 index cards	• 10 spring clothespins
• 32 cm piece of white paper	• small binder clip	• flashlight
	• hole punch	

Steps:

1. Using a metric ruler, draw a line lengthwise down the center of a sheet of white paper.
2. Mark the line at 6 cm, 12 cm, 18 cm, 24 cm, and 32 cm.
3. Find the center of an index card by drawing a diagonal line from one corner of the card to the opposite corner of the card. Draw a second line connecting the two other corners.
4. Align this index card on top of three other index cards. Secure the four cards with a binder clip.
5. Punch a hole in the center of all four cards.
6. Clamp two spring clothespins on either side of the short end of an index card. Repeat this step for the four remaining index cards.
7. Using the clothespins as a base, position the four index cards with holes on the 6 cm, 12 cm, 18 cm, and 24 cm marks. Make sure that the holes in the cards are aligned.
8. Place the fifth index card without the hole on the 32 cm mark.
9. Turn off the classroom lights.
10. Point the flashlight toward the holes in the cards and turn on the flashlight. Observe the path of the light.
11. Adjust the second index card so the hole is no longer in line with the others. Observe the path of the light.

ANALYZE AND CONCLUDE:

1. Describe the appearance of light on the cards when the holes were lined up.

____

____

____

2. Describe the appearance of light on the cards when the holes were no longer aligned in Step 12. ______________________________

____

Lab 6.3.2A The Path of Light

3. What circle of light is the most intense? Why? ____________________

____

____

4. What path does light travel? ____________________

5. Explain why light was not visible on the third card when the holes were no longer aligned. ____________________

____

Name: ______________________________ Date: ____________

Lab 6.3.3A Color Combinations

QUESTION: How is the color of an object determined?

HYPOTHESIS: ______________________________

____

EXPERIMENT:

You will need:	• pin	• red, blue, and green paint
• scissors	• 3 paintbrushes	• pencil
• cardboard		

Steps:

1. Cut out four 12 cm diameter cardboard disks.
2. Using a pin, puncture a hole in the center of each disk.
3. Refer to the templates on the following page. Paint half of the first disk red and the other half blue. Allow the paint to dry.
4. On the second disk, paint alternating segments (pie slices) blue and green. Allow the paint to dry.
5. Paint half of the third disk green and the other half red. Allow the paint to dry.
6. On the fourth disk, paint one third of the disk red, one third blue, and one third green. Allow the paint to dry.
7. When the disks are dry, insert a pin through the hole of each disk and spin the disk rapidly.
8. Record your observations, including the colors seen when the disks spin.

____

____

____

____

9. Fill in the colors of the disk below from Step 6 and indicate which three primary colors of light are reflected and absorbed in each pigment.

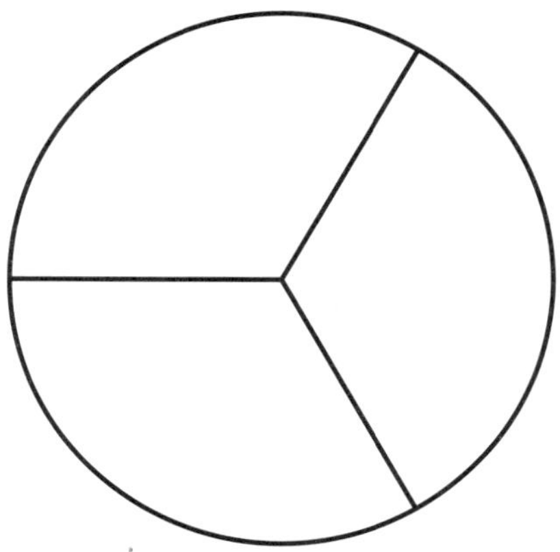

Lab 6.3.3A Color Combinations

ANALYZE AND CONCLUDE:

1. What are the primary colors of light? ______________________

____

2. Why did your eyes observe different colors when you twirled the disks? __________

____

____

____

____

____

3. What color does the fourth disk appear? Why? ______________________

____

____

1.

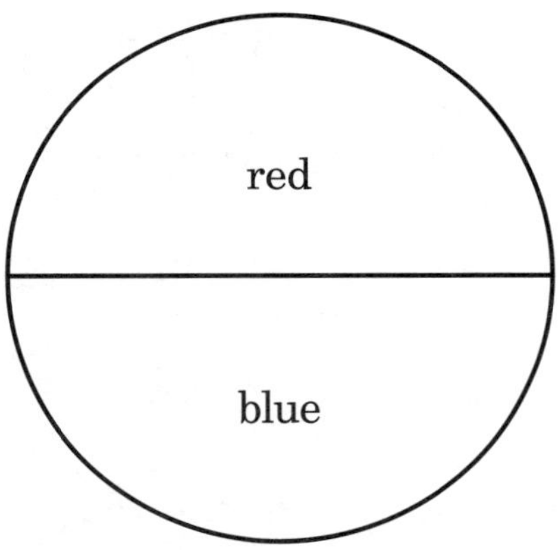

2.

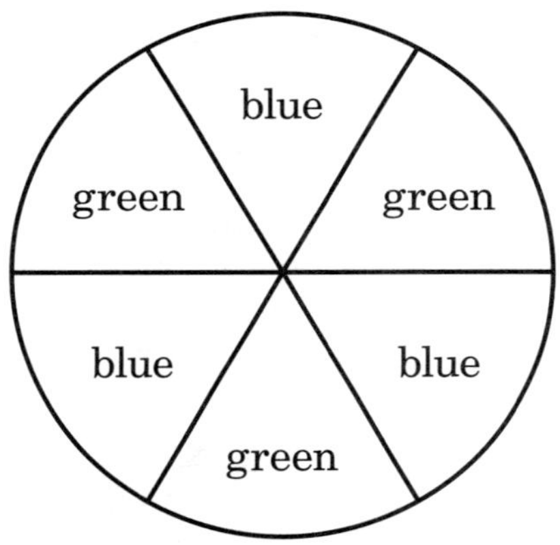

3.

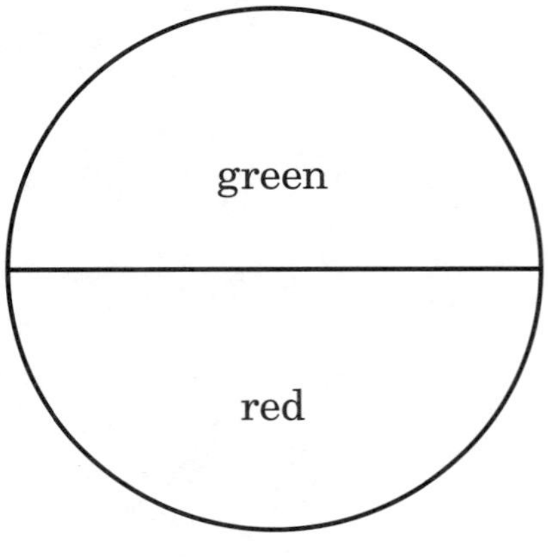

4.

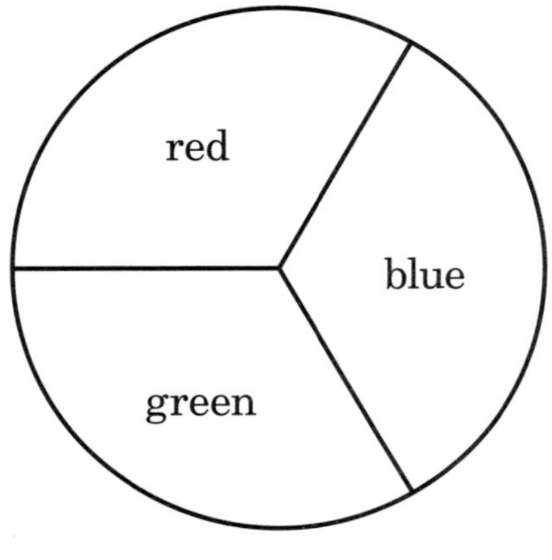

Name: ______________________ Date: __________

Lab 6.3.6A Reflection

QUESTION: How are light waves reflected on a smooth surface?

HYPOTHESIS: ______________________

____

EXPERIMENT:

You will need: • 2 flat mirrors	• protractor	• candle

Steps:

1. Place two flat mirrors side by side along the straight edge of a protractor.
2. Position a candle in front of the two mirrors and observe the number of candles that are visible in the reflection of the mirrors.
3. Maintain contact between the two edges of the mirrors as you move one mirror face toward the opposite mirror until three candle images are observed.
4. Adjust the mirror to determine the maximum and minimum angle needed to produce three images. Record the angle range in the table below.
5. Repeat Steps 3–4 to determine the maximum and minimum angles needed to produce 4–7 candle images. Record the angle ranges in the table below.

Number of Candle Images	Minimum Angle	Maximum Angle
2		
3		
4		
5		
6		
7		

ANALYZE AND CONCLUDE:

1. How did the number of images relate to the angle of the mirrors? ______________________

____

2. Hypothesize the number of candles that would appear if the mirror faces were placed parallel to one another. ______________________

____

3. Did the mirrors demonstrate regular or diffuse reflection? Explain. ______________________

____

____

____

Name: ______________________________ Date: ______________

Lab 6.3.7A Artificial Lights

QUESTION: Does an incandescent light bulb or fluorescent light bulb emit more heat? Why?

HYPOTHESIS: ______________________________

____

EXPERIMENT:

You will need: • 3 incandescent light bulbs of various power ratings	• screw • paper cup • thermometer	• 2 fluorescent light bulbs of various power ratings • LED light bulb

Steps:

1. Turn on an incandescent light bulb and allow the bulb to remain undisturbed for 10 minutes.
2. Poke a small hole with a screw through the base of a paper cup.
3. Insert the thermometer through the hole until the thermometer tip is positioned in the center of the paper cup.
4. Record the initial temperature: ________
5. While holding the free end of the thermometer, position the opening of the cup 10 cm away from the light source.
6. Keep the thermometer and cup in place for three minutes and record the temperature in the table below.
7. Allow the thermometer to cool to the initial temperature.
8. Repeat Steps 1–7 for the other incandescent, fluorescent, and LED light bulbs.

Type of Light Bulb	Power Rating (W)	Temperature (°C)
Incandescent #1		
Incandescent #2		
Incandescent #3		
Fluorescent #1		
Fluorescent #2		
LED		

ANALYZE AND CONCLUDE:

1. Which type of light bulb generated the most amount of heat? ______________________________

____

2. Which type of light bulb generated the least amount of heat? ______________________________

____

Lab 6.3.7A Artificial Lights

3. Describe how an incandescent bulb produces light. ______________________

4. Describe how a fluorescent bulb produces light. ______________________

5. What are some advantages and disadvantages of using both types of bulbs?

Name: ____ Date: ________________

Emission Spectrums

WS 6.3.1A

Using colored pencils, fill in the emission spectrum for each light source. If a continuous spectrum is visible, write *continuous spectrum* in the space provided.

Incandescent light bulb

Compact fluorescent lamp (CFL)

Light-emitting diode lamp (LED)

Insect light

UV lamp

Home light source 1

Home light source 2

Name: ______________________ Date: ____________

The Electromagnetic Spectrum

WS 6.3.1B

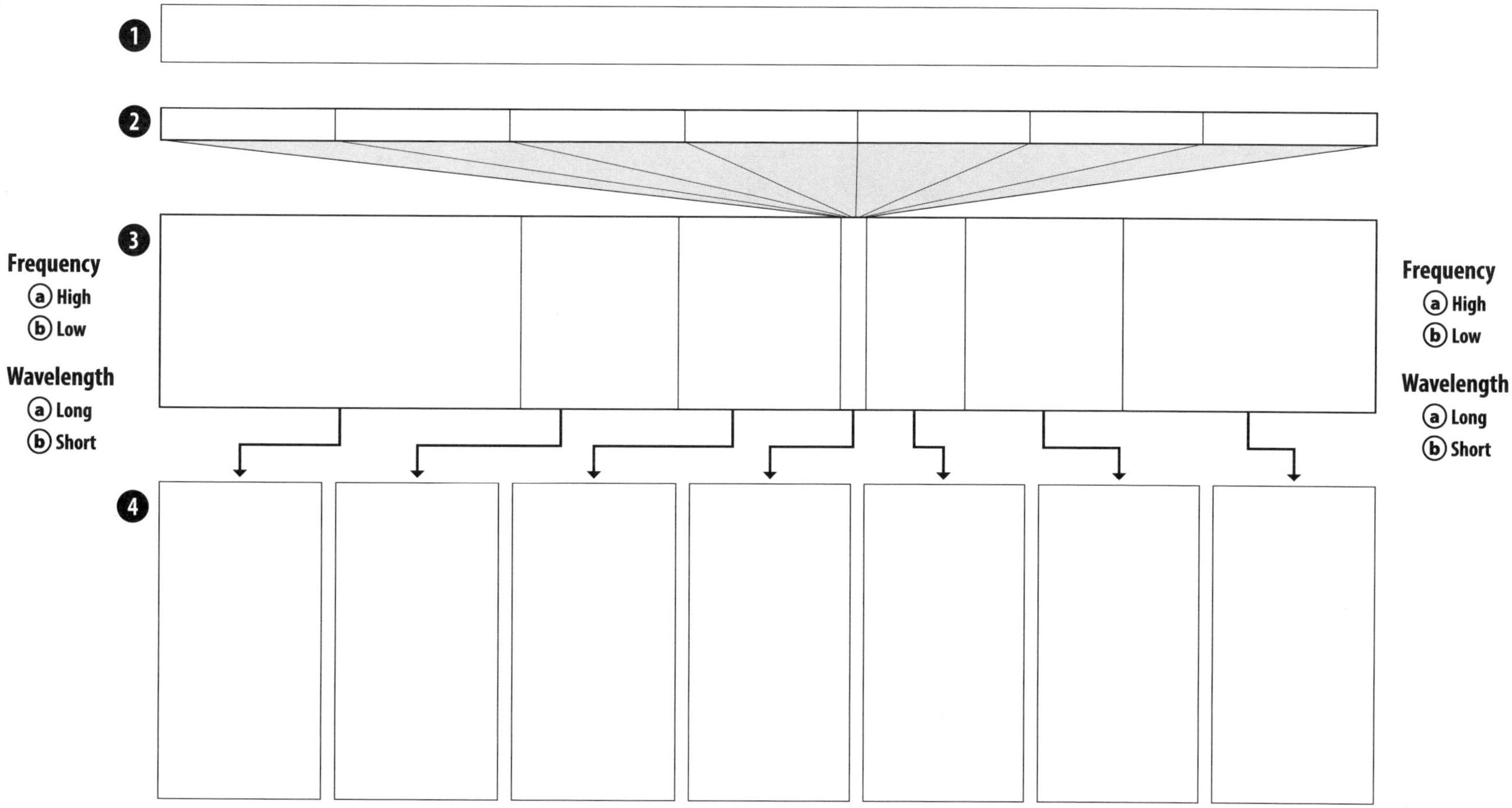

1. In the box, draw an electromagnetic transverse wave that illustrates the frequency and wavelength of each region.
2. Use colored pencils to apply the correct color to the visible light boxes.
3. Fill in the blank spectrum with the name of each type of radiation. Fill in the answer for the frequency and wavelength at each end of the electromagnetic spectrum.
4. Research different uses of each type of radiation and fill in the boxes with your answers.

Name: ______________________ Date: ____________

Lab 6.4.1A Reflecting Reflections

QUESTION: How many reflections can be seen with two plane mirrors?

HYPOTHESIS: ______________________

____

EXPERIMENT:

You will need:		
• 2 plane mirrors	• masking tape	• protractor
	• plain paper	• paper clip

Steps:

1. Place one mirror on top of the other with the reflected surfaces facing inward. Tape the mirrors together on one long side so the mirrors will open and close.
2. Open the mirrors and use tape to label the left mirror *L* and the right mirror *R*.
3. Stand the mirrors on a sheet of paper. Close the mirrors to an angle of 72°.
4. Bend one end of the paper clip at a 90° angle. Place it near the surface of the right mirror.
5. Count the number of images of the paper clip in both the left and right mirrors. Record the number.
6. The two mirrors create an image of a circle divided into wedges. How many wedges do you see? Record the number.
7. Open the mirrors to a 90° angle. Count and record the images of the paper clip and the number of wedges in the circle.
8. Repeat Step 7 using a 120° angle.

Angle	Number of Images R	Number of Images L	Number of Wedges
72°			
90°			
120°			

ANALYZE AND CONCLUDE:

1. What is the relationship between the number of wedges and paper clip images that can be seen? ______________________

____

2. Using this information, what angle would create six images? ______________________

____

3. How many images would be produced if two plane mirrors were directly opposite each other? ______________________

____

4. What type of images are formed? ______________________

Name: ____________________ Date: ____________

Lab 6.4.2A Focal Lengths

QUESTION: How does the curve of a convex lens affect the focal length?

HYPOTHESIS: ____________________

____

EXPERIMENT:

You will need:	• paper	• meterstick
• various hand lenses		

Steps:

1. Place a hand lens in direct sunlight above a sheet of white paper.
2. Move the lens back and forth until the dot of light is smallest. This is the focal point of the lens.
3. Keep the lens in that position until the paper starts to burn.
4. Use a meterstick to measure the focal length.
5. Repeat the procedure with the other hand lenses.

Hand Lens	Focal Length

ANALYZE AND CONCLUDE:

1. Which lens had the longest focal length? ____________________

2. Which lens had the shortest focal length? ____________________

3. What is the relationship between the curvature of the lens and the focal length?

____

____

4. What would happen if the experiment was repeated with concave lenses? Why?

____

____

____

Name: ______________________ Date: ____________

Lab 6.4.2B Convex Lens

QUESTION: What type of image does a convex lens produce?

HYPOTHESIS: ______________________

____

EXPERIMENT:

You will need:	• transparent tape	• large hand lens
• black paper	• utility knife	• white poster board

Steps:

1. Tape black paper to the windows to block out the light.
2. Cut a slit in the center of one piece of paper to allow a narrow beam of sunlight through.
3. Turn out the lights.
4. Position a hand lens in the sunbeam, and place white poster board behind the hand lens.
5. Adjust the poster board and lens until a focused image is produced.
6. Turn the lights back on.
7. Place printed text on a table and use the hand lens to view it.

ANALYZE AND CONCLUDE:

1. Was the image on the poster board a real or virtual image? Why? ______________________

____

____

2. Was the image of the written text a real or virtual image? Why? ______________________

____

____

3. What kind of lens is a hand lens? ______________________

4. What kind of image is produced when the focal point is between the object and the hand lens? ______________________

____

____

Name: ______________________________ Date: ______________

Lab 6.4.3A Images

QUESTION: Why do different optical instruments use different focal lengths?

HYPOTHESIS: ______________________________

____

EXPERIMENT:

You will need:		
• glass test tube	• test tube stopper	• metric ruler
	• index card	

Steps:

1. Fill the test tube with water and fasten the stopper securely.
2. Print the words *carbon dioxide* on the index card.
3. Place the card on the table and position the test tube 0 cm, 1 cm, 2 cm, and 10 cm away from the card. Look at the word through the test tube at each position.
4. Observe the changes in the image produced at these different distances.

ANALYZE AND CONCLUDE:

1. What did you see at 0 cm? ______________________________

2. What kind of optical instrument acts like this? ______________________________

3. How did the image look at 1 cm? ______________________________

____

4. Which distance was near the focal point? ______________________________

5. What did you see at 2 cm? ______________________________

____

6. At 2 cm, how many focal lengths was the object from the test tube? ______________

____

7. At 2 cm, what kind of image did the test tube produce, real or virtual? ______________

____

8. What kinds of optical instruments use this type of image? ______________________________

____

9. What did you see at 10 cm? ______________________________

____

10. At 10 cm, what kind of image did the test tube produce, real or virtual? ______________

____

Name: ______________________________ Date: ______________

Lab 6.4.3B Telescope

QUESTION: How does a telescope work?

HYPOTHESIS: ______________________________

____

____

EXPERIMENT:

You will need:
- small double-convex lens with a focal length of about 2.5 cm
- large plano-convex lens with a focal length of about 20 cm
- meterstick
- 2 mailing tubes, at least 60 cm in length
- scissors
- tape
- cardboard
- craft glue

Steps:

1. Hold the small lens (eyepiece) close to your eye and the large lens (objective lens) at arm's length. Look through the lenses at an object across the room. Bring the larger lens closer until the image created by the lenses is focused.
2. Have a partner measure the distance between the two lenses. Record the distance. ______________________________
3. Cut a mailing tube in half lengthwise. Cut off and discard a 1 cm strip down its length. Squeeze the cut tube together to form a slightly smaller tube than the original. Tape it securely, ensuring that this tube slides easily into the uncut mailing tube.

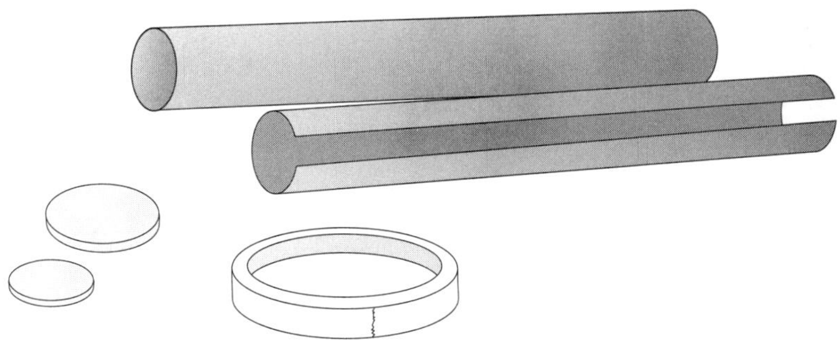

4. Cut two rings of cardboard with an outer diameter the diameter of the smaller tube and an inner diameter slightly smaller than the diameter of the smaller lens. Then cut two rings with an outer diameter the diameter of the larger tube and an inner diameter slightly smaller than the diameter of the larger lens.
5. Glue the eyepiece lens between the smaller rings and the objective lens between the larger rings.
6. Tape the lenses to the tube sections. Insert the smaller tube into the larger tube so the lenses are on opposite ends.
7. Look through the eyepiece, and move the smaller tube in and out to magnify and focus on distant objects. WARNING: Never look directly at the sun, with or without a telescope.

Lab 6.4.3B Telescope

ANALYZE AND CONCLUDE:

1. Is the telescope a reflecting or refracting telescope? ____________________

2. Why does the telescope use two lenses? ____________________

____

3. Would a third lens further enlarge the image? Why? ____________________

____

____

____

4. Label the parts of the telescope and diagram the light rays.

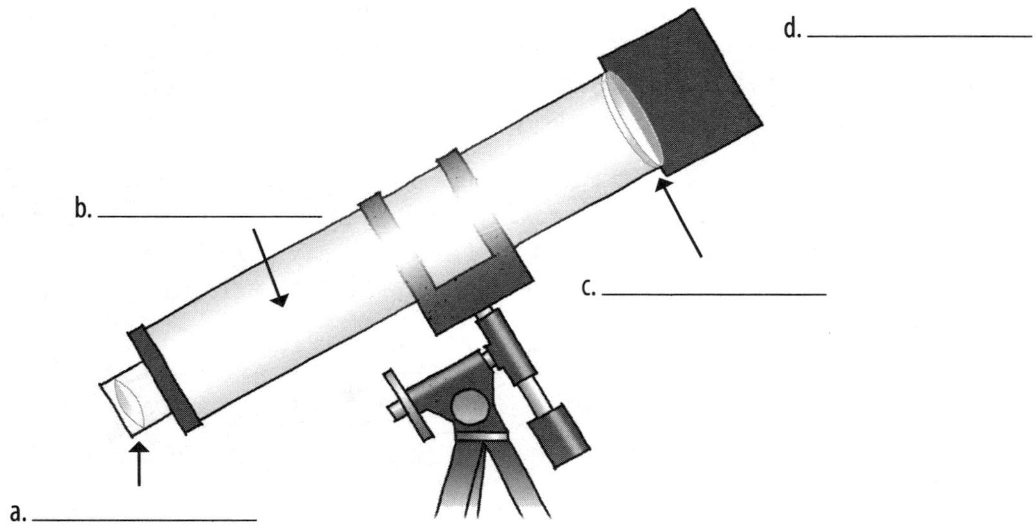

Name: ______________________________ Date: ____________

Lab 6.4.4A Polarized Syrup

QUESTION: What happens to polarized light when it travels through syrup?

HYPOTHESIS: ______________________________

____

EXPERIMENT:

You will need:	• 2 pieces of polarizing film	• bright light source, such as a desk lamp or flashlight
• 100 mL clear, graduated cylinder	• tape	
• light corn syrup	• white paper	

Steps:

1. Fill the graduated cylinder with 40 mL of syrup.
2. Tape one piece of polarizing film to the bottom of the graduated cylinder.
3. Place the white paper on a table.
4. Shine the light source on the white paper.
5. Have a partner hold the cylinder above the reflection of the light.
6. Look down through the other polarizing film into the cylinder. Slowly rotate the top filter and take note of the color changes in the syrup.
7. Have a partner add 30 mL of syrup as you continue to look through the top filter. Note any changes.

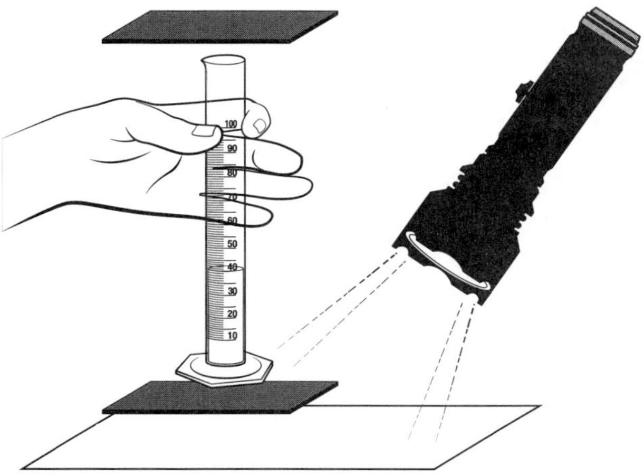

ANALYZE AND CONCLUDE:

1. What happened when the polarizing film was turned? ______________________________

____

2. What happened when more corn syrup was added? ______________________________

____

3. Why did the polarizing film show different colors in the corn syrup? ______________________________

____

____

____

____

Name: ____________________ Date: ____________

Light Ray Diagrams

WS 6.4.1A

Create ray diagrams to demonstrate the location and type of image formed in each case.

1. Demonstrate a concave mirror reflecting an object positioned behind the focal point. Draw a light ray from the top of the object to the mirror and parallel to the optical axis. Draw a second light ray from the top of the image through the focal point to the surface of the mirror. Now use the law of reflection to create the reflected rays and draw the image created. Is it real or virtual? ____________________

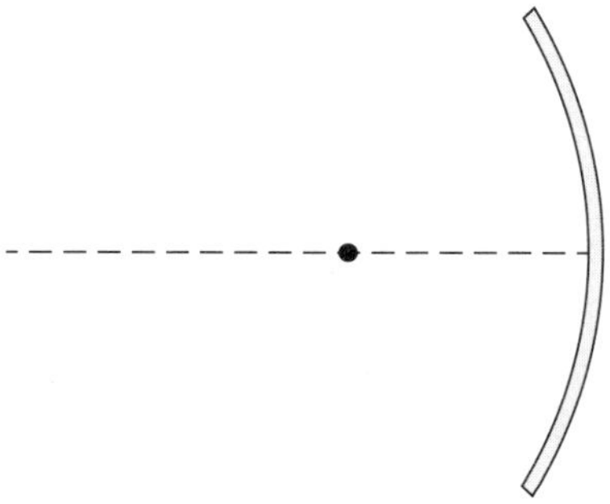

2. Show how a concave mirror is used in a flashlight. Draw light rays from the focal point to the surface of the mirror at 30° and 45° above the optical axis and 30° and 45° below the optical axis. Draw light rays of the angles of reflection that are parallel to the optical axis.

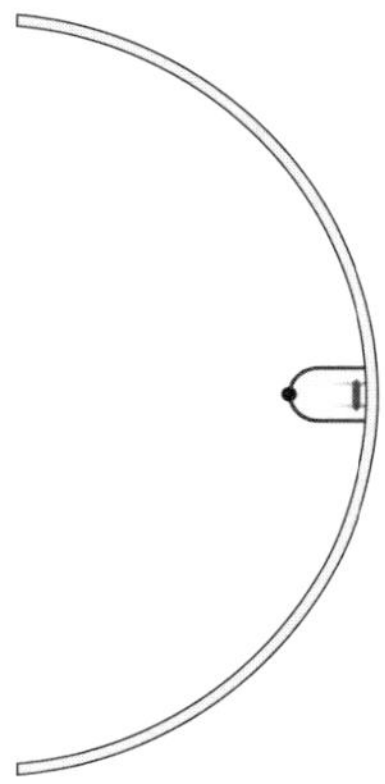

3. Demonstrate the location and type of image formed when an object is placed between the focal point and the surface of a concave lens. Draw a light ray from the top of the object to the surface of the mirror that is parallel to the optical axis. Draw another ray from the top of the object 45° from the parallel ray. Use the law of reflection to indicate the light rays resulting from the angles of incidence. Extend those rays behind the mirror until they converge. The point where the lines appear to converge is the top of the image. Draw the image. Is it real or virtual? How does the size relate to the object?

____

____

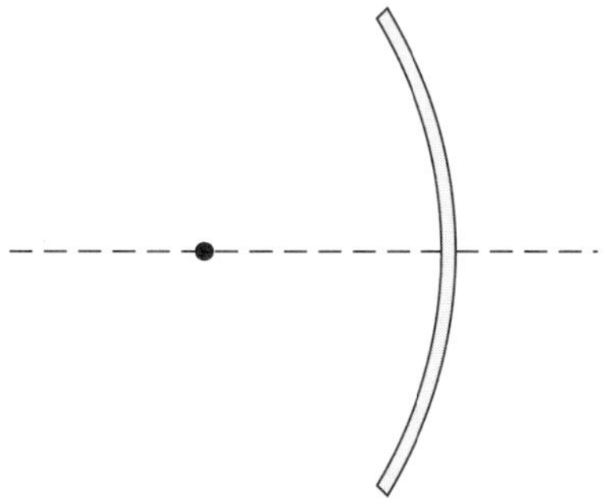

Light Ray Diagrams

4. Create a light ray diagram demonstrating the image created when an object is placed in front of a convex mirror. Draw a light ray from the top of the object to the surface of the mirror that is parallel to the optical axis. Draw another ray from the top of the object that extends 30° below the first ray to the surface of the mirror. Use the law of reflection to create rays demonstrating the angles of incidence. Extend those lines behind the mirror. Where the lines appear to converge is at the top of the image. Draw the image. What type of image is it? How does the size relate to the object?

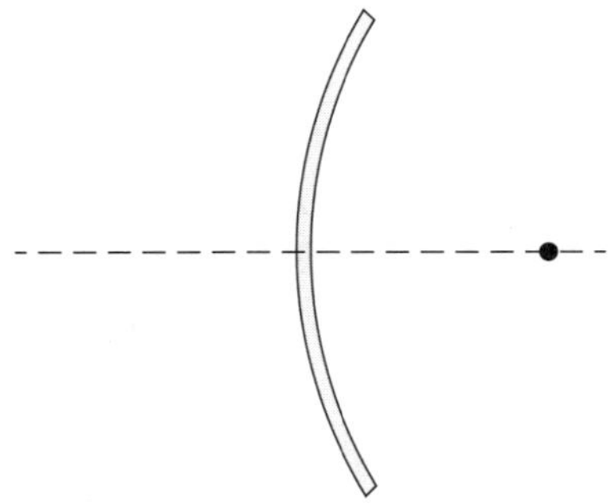

Name: ______________________________ Date: ______________

Lab 7.1.1A Induction

QUESTION: How do objects obtain a charge?

HYPOTHESIS: ______________________________

____

EXPERIMENT:

You will need:		
• support rod	• support stand	• glass or rubber rod
• clamp	• 15 cm piece of string	• wool
	• pith ball	• silk

Steps:

1. Secure a support rod with a clamp to the upper portion of a support stand.
2. Tie one end of the string to a pith ball and the other end of the string to the support rod.
3. Rub a glass or rubber rod with a piece of wool for several seconds.
4. Bring the rod near the pith ball for about 30 seconds, but do not touch the ball with the rod.
5. Record your observations. ______________________________

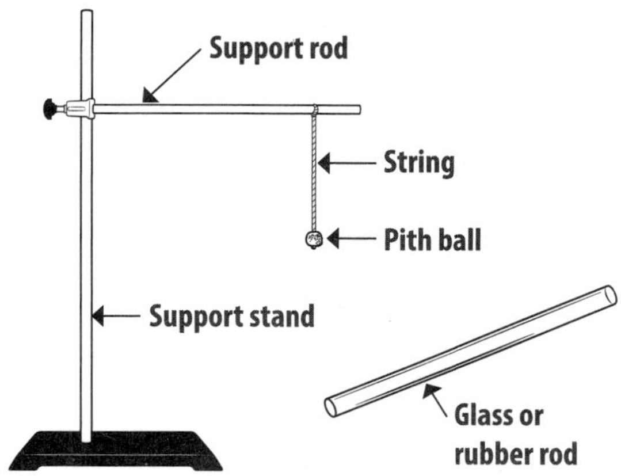

____

____

____

____

6. Discharge the ball by touching the back of your hand to the pith ball.
7. Rub the glass or rubber rod again with a piece of wool for several seconds.
8. Touch the rod to the pith ball and release it, doing both without touching the pith ball with your hand.
9. Bring the rod near the pith ball. Record your observations. ______________________________

____

10. Rub the glass or rubber rod with a piece of silk.
11. Bring the rod near the pith ball. Record your observations. ______________________________

____

____

Lab 7.1.1A Induction

ANALYZE AND CONCLUDE:

1. Explain your observations in Step 5. ______

______

______

______

______

______

2. Why was there a dramatic result in Step 9? ______

______

______

3. Was the charge on the rod in Step 7 the same as the charge on the rod in Step 10? Explain. ______

______

______

______

______

______

______

______

Name: ______________________________ Date: ______________

Lab 7.1.1B Charge Detection

QUESTION: How does an electroscope show that a charge is present?

HYPOTHESIS: ______________________________

____

EXPERIMENT:

You will need:	• aluminum foil	• glass or rubber rod
• large paper clip	• mason jar with lid	• wool
• scissors	• nail	• rubber stopper
• metric ruler	• tape or clay	• copper wire

Steps:

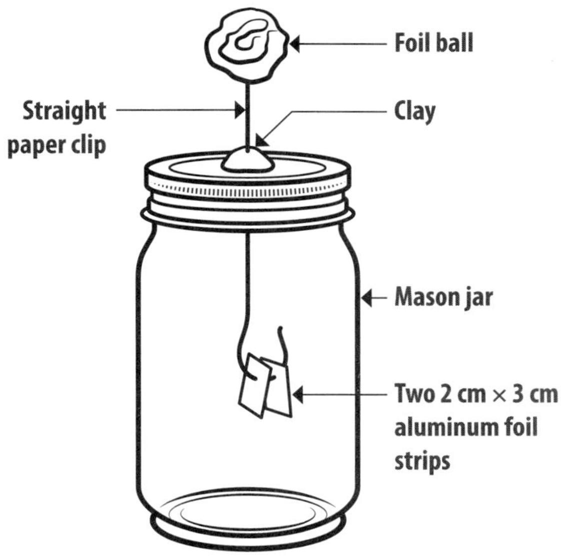

1. Straighten a large paper clip most of the way, leaving one end of the paper clip curved.
2. Cut out two 2 cm × 3 cm pieces of aluminum foil.
3. Hang the pieces of aluminum foil on the curved end of the paper clip.
4. Puncture the lid of a mason jar with the tip of a nail and insert the pointed end of the paper clip through the hole so the curved end of the paper clip and foil pieces are inside the jar.
5. Secure the paper clip with a piece of tape or clay. The pieces of aluminum foil should hang freely without touching the bottom or sides of the jar.
6. Close the lid on the jar.
7. Crumple a piece of aluminum foil to form a ball with a 4 cm diameter.
8. Insert the pointed end of the paper clip halfway into the center of the foil ball.
9. Rub a glass or rubber rod with a piece of wool for several seconds.
10. Touch the foil ball with the rod.
11. Record your observations. ______________________________

    ____

12. Repeat this procedure using a rubber stopper or a copper wire in place of the foil ball.
13. Record your observations. ______________________________

    ____

    ____

Lab 7.1.1B Charge Detection

ANALYZE AND CONCLUDE:

1. Explain your observations in Step 10. ______________________________

____

____

2. Is aluminum foil, a rubber stopper, or copper wire a better transmitter of electric charge? ______________________________

____

____

3. How can you determine if the leaves of an electroscope have a positive or negative charge? ______________________________

____

____

____

Name: ______________________________ Date: ______________

Lab 7.1.2A Conductors and Insulators

QUESTION: What types of materials are the best conductors?

HYPOTHESIS: ______________________________

____

EXPERIMENT:

You will need:
- D cell battery
- battery holder
- 3.7 V miniature light bulb
- light bulb socket
- 3 pieces of insulated wire with alligator clips

Steps:

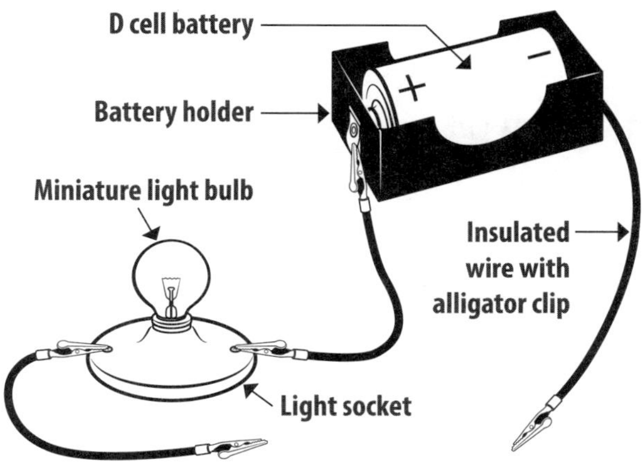

1. Place the battery in the battery holder and the light bulb in the light socket.
2. Attach the end of one wire to the positive battery terminal and the other end of the wire to a connection on the light socket.
3. Attach the end of the second wire to the other connection on the light socket. Leave the other end of the wire free.
4. Attach the end of the third wire to the negative battery terminal. Leave the other end of the wire free. Do not allow the two free wire ends to touch.
5. At the same time, touch the free ends of the two wires to various objects. Record your results in the table below.

Material	Light bulb glows? (yes or no)	Conductor or insulator?

Lab 7.1.2A Conductors and Insulators

ANALYZE AND CONCLUDE:

1. What materials are conductors? ______

2. What materials are insulators? ______

3. Are metals or nonmetals the best conductors? Why? ______

4. Did all the conductors cause the bulb to illuminate with the same intensity? Why? ______

Name: ______________________ Date: ____________

Lab 7.1.3A Air Battery

QUESTION: How is an electric current conducted through a dry cell battery?

HYPOTHESIS: ______________________

EXPERIMENT:

You will need:		
• paper towels	• 75 mL saturated sodium chloride solution	• graphite rod
• 20 cm × 20 cm sheet of aluminum foil	• pipette	• spatula
• 100 mL beaker	• 2 insulated wires with alligator clips, one red wire and one black wire	• 10 g activated charcoal
		• multimeter

Steps:

1. Dry all laboratory surfaces before conducting this experiment.
2. Place the aluminum foil sheet flat on the surface of the table.
3. Position a paper towel on top of the sheet of foil.
4. Fill a 100 mL beaker with 75 mL of sodium chloride solution.
5. Using a pipette, moisten the entire paper towel with the solution.
6. Attach the alligator clip of the red insulated wire to the graphite rod.
7. Position the free end of the graphite rod in the center of the paper towel.
8. Using a spatula, scoop about 10 g of charcoal onto the paper towel to cover the free end of the graphite rod.
9. Fold the aluminum foil and paper towel in half so the length of the graphite rod is in the crease of the fold. Ensure the paper towel remains between the charcoal and foil.
10. Attach the alligator clip of the black insulated wire to the aluminum foil that is opposite where the red insulated wire extends from the foil.
11. Using a multimeter, measure and record the voltage and current.
    - Voltage: ________
    - Current: ________

ANALYZE AND CONCLUDE:

1. If this battery was connected to a light bulb, would the voltage or the current cause the bulb to illuminate? ______________________

2. What created a current in this battery? ______________________

3. What served as the electrolyte in this battery? ______________________

4. What type of battery cell did you create? ______________________

Name: ______________________________ Date: ______________

Lab 7.1.4A Series Circuits

QUESTION: How does current flow in a series circuit?

HYPOTHESIS: ______________________________

____

EXPERIMENT:

You will need:	• 6 V battery	• flat-head screwdriver
• two 20 cm pieces of insulated wire	• 3 miniature light bulbs and sockets	• two 6 cm pieces of insulated wire

Steps:

1. Attach one end of a 20 cm piece of insulated wire to the positive terminal of a 6 V battery.
2. Attach the end of a second piece of 20 cm insulated wire to the negative terminal of a 6 V battery. Do not allow the free wire ends to touch.
3. Using a flat-head screwdriver, connect three miniature light bulb sockets with two pieces of 6 cm insulated wire.
4. Attach the end of the wire that is connected to the positive terminal of the battery to one of the free light bulb sockets.
5. Attach the end of the wire that is connected to the negative terminal of the battery to the remaining free light bulb socket.

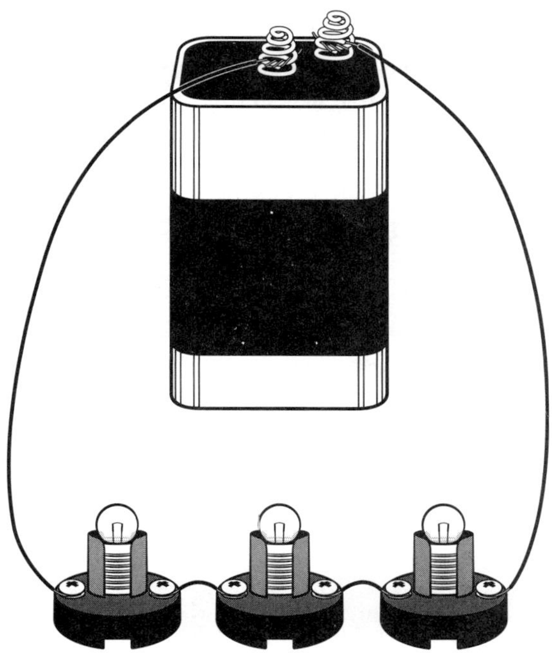

6. Record your observations. ______________________________

   ____

7. Remove one light bulb from its socket. Record your observations. ______________

   ____

8. The circuit should remain connected for a short period of time, long enough to make observations. Disconnect the circuit immediately after observations are complete.

ANALYZE AND CONCLUDE:

1. Why did all of the light bulbs illuminate in Step 6? ______________

____

____

Lab 7.1.4A Series Circuits

2. Explain your observations in Step 7. ______________________________

____

____

____

____

3. Name examples of devices that use series circuits. ______________________________

____

____

4. Hypothesize whether the bulbs would be brighter if only two light bulbs were included in the series circuit rather than three bulbs. Explain your answer.

____

____

____

____

____

____

Name: ______________________ Date: __________

Lab 7.1.4B Parallel Circuits

QUESTION: How does current flow in a parallel circuit?

HYPOTHESIS: ______________________

____

EXPERIMENT:

You will need:	• 6 V battery	• 3 miniature light bulbs and sockets
• two 20 cm pieces of insulated wire	• four 6 cm pieces of insulated wire	• flat-head screwdriver

Steps:

1. Attach one end of a 20 cm piece of insulated wire to the positive terminal of a 6 V battery.
2. Attach the end of a second piece of 20 cm insulated wire to the negative terminal of a 6 V battery. Do not allow the free wire ends to touch.
3. Obtain four pieces of 6 cm insulated wire and three miniature light bulbs in sockets.
4. Using a flat-head screwdriver, connect the first two lights with two pieces of insulated wire.
5. Connect the second and third light with an additional two pieces of insulated wire.
6. Attach the free end of one of the wires that is connected to the battery to one of the screws in the first light bulb socket.
7. Attach the free end of the second wire that is connected to the battery to the second screw in the first light bulb socket.

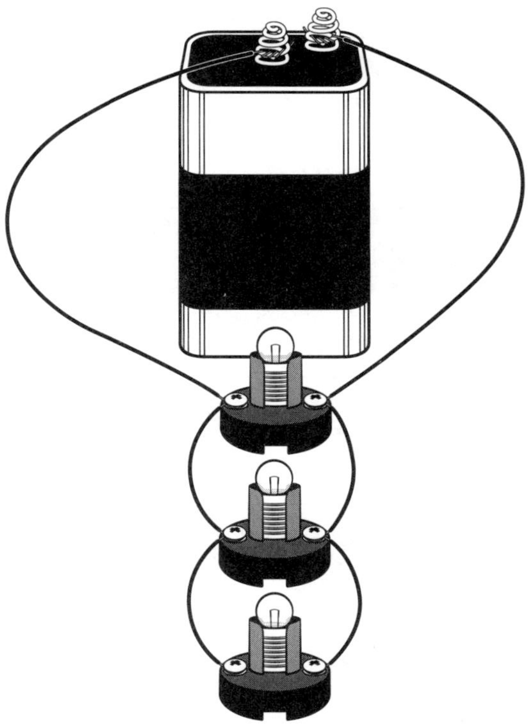

8. Record your observations. ______________________

____

9. Remove two of the light bulbs from their sockets. Record your observations.

____

____

ANALYZE AND CONCLUDE:

1. Why did all of the light bulbs illuminate in Step 8? ______________________

____

Lab 7.1.4B Parallel Circuits

2. Explain your observations in Step 9. ______

3. Are bulbs brighter in series circuits or parallel circuits? Explain. ______

4. Are electrical systems in homes series or parallel circuits? Explain. ______

Name: ____________________ Date: __________

Voltage, Resistance, and Current WS 7.1.5A

1. The filament in an incandescent light bulb uses a current of 0.5 A when the voltage is 60 V. What is the resistance of the filament?

2. The resistance of a wire is 8 ohms. It carries a current of 4 amps. What is its voltage?

3. The resistance of a wire is 40 Ω. Its voltage is 100 volts. What current does it carry?

4. What is the resistance of a wire that uses a current of 6 amps when the voltage is 90 V?

5. What current does a wire carry when its resistance is 70 ohms and its voltage is 525 V?

6. What is the voltage of a wire with a resistance of 6.5 ohms and a current of 1.5 A?

7. A current of 4.2 amps runs through a circuit with a resistance of 20 ohms. What is the voltage?

8. If a battery in a circuit is 25 V and the resistance is 15 ohms, what is the current?

9. Given a resistance of 1,200 Ω and a current of 0.05 amps, what is the voltage?

10. Battery potential is 40 ohms and voltage is 13 V. Calculate I.

Name: ______________________________ Date: ______________

Electrical Power and Energy **WS 7.1.6A**

Use the following equations to solve these questions:

$$I = V \div R$$
$$P = V \times I$$
$$E = P \times t$$

1. A television uses 1.5 amps when it is connected to a voltage of 120 V. How much electrical power does it use?

2. A hair dryer has a power rating of 1,500 W. How much electrical energy does it use in 30 seconds?

3. How much electrical energy would five 60 W light bulbs consume if they were left on for six hours?

4. Several 100-watt bulbs burned for eight hours and consumed four kilowatt hours of electrical energy. How many bulbs were burning?

5. What is the current when a 60 W lamp is connected to 120 V?

6. If part of an electric circuit dissipates energy at 10 W when it draws a current of 5 A, what voltage is impressed across it?

7. Calculate the power of the light bulb in the circuit shown.

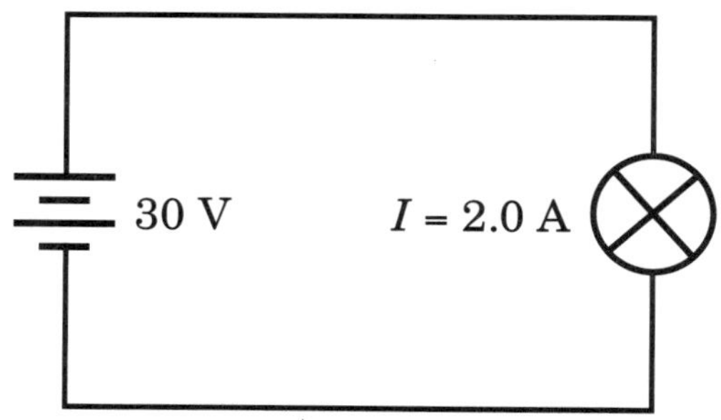

Name: ____________________________________ Date: ______________

Lab 7.2.1A Magnetic Strength

QUESTION: What is the effect of using more magnets in a magnetic field?

HYPOTHESIS: ____

____

EXPERIMENT:

You will need:	• 15 cm length of sewing thread	• 4 identical disk magnets
• heavy cardboard, 30 cm × 20 cm	• small iron nail	
• tape	• tack	
• protractor	• metric ruler	

Steps:

1. Attach the cardboard to the wall. Tape the protractor to one end of the board so the straight end is parallel to the edge of the cardboard and 1 cm from the edge.
2. Tie one end of the thread around the head of the nail. Tack the other end of the thread to the cardboard just above the straight edge of the protractor at the 0° point.
3. Tape the metric ruler to the cardboard just below the suspended point of the nail. Align the ruler parallel to the straight edge of the protractor and with the 0 cm mark at the point of the nail.

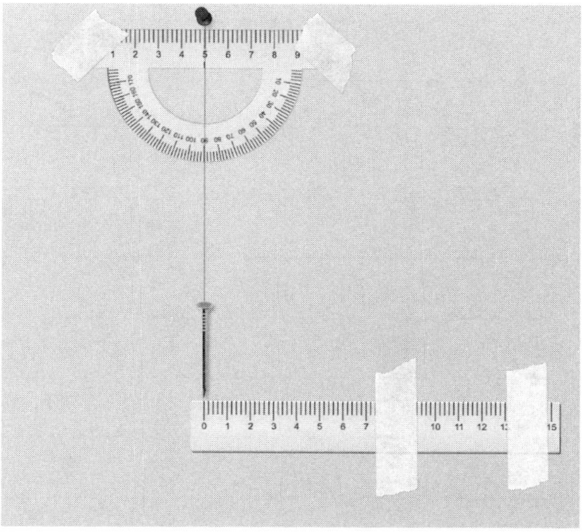

4. Hypothesize the distance at which one magnet will first cause the nail to move. Slide the magnet slowly from the far end of the ruler toward the nail. Record the distance the nail first moves toward the magnet. Repeat the process two more times and calculate the average. Record the data in the table below.
5. Follow the same procedure using two, three, and then four magnets together.

Distance to Magnetic Force

# of Magnets	Hypothesis	Trial 1	Trial 2	Trial 3	Average
1 magnet					
2 magnets					
3 magnets					
4 magnets					

Lab 7.2.1A Magnetic Strength

ANALYZE AND CONCLUDE:

1. Why is it important to conduct several trials and take an average? ______________

____

____

2. Did your hypotheses become more accurate with each trial? Why? ______________

____

____

____

3. What happened when the number of magnets was increased? ______________

____

____

4. Is the repulsion force of magnets as strong as the attractive force? Design an experiment to test your answer. ______________

____

____

____

____

____

____

____

Name: ______________________ Date: __________

Lab 7.2.2A Electromagnetism

QUESTION: How can an electric current generate a magnetic field?

HYPOTHESIS: ______________________

____

EXPERIMENT:

You will need:		
• 1 m length of insulated copper wire	• large iron nail	• 6 V battery
	• 12 small paper clips	

Steps:

1. Wrap the center portion of the wire around the iron nail. Make five tight, even loops around the nail. Do not overlap the loops.
2. Hold the head of the nail while a lab partner connects the two ends of the wire to the two battery terminals. Caution: The electric current will produce heat in the wire, indirectly heating the nail. Disconnect the wires if the nail becomes too hot to handle.
3. Use the nail as a magnet. Determine how many paper clips the nail is able to pick up. For a more accurate assessment of the magnetic force, break the paper clips in half and then count the number of halves the nail can pick up.
4. Create a data chart, and record the number of paper clips or half paper clips the magnet is able to pick up.
5. Repeat this procedure making 10, 15, and 20 loops of wire around the nail. Record the data on the chart.
6. Disconnect the wires.

ANALYZE AND CONCLUDE:

1. Why was the nail magnetized? ______________________

____

____

2. What happened when more coils of wire were looped around the nail? Why? ______

____

____

____

3. What else might change the magnetic strength of the nail? ______________________

____

____

____

Name: ______________________________ Date: ______________

Lab 7.2.2B Electromagnetic Induction

QUESTION: How can a magnet be used to create an electric current?

HYPOTHESIS: ______________________________

____

EXPERIMENT:

You will need:		
• 1 m length of insulated copper wire	• compass	• strong bar magnet
	• 3 m length of insulated copper wire	

Steps:

1. Wrap the 1 m length of copper wire around the compass 20–25 times, leaving several centimeters of each end of the wire straight.
2. Place the compass on a table. Align its needle to point in the same direction as the coils of wire around it.
3. Starting from the center of the 3 m length of copper wire, loop the wire around four fingers 15 times.
4. Connect the two ends of the 3 m length of wire to the two ends of the 1 m wire by tying them like a twist tie.
5. Move the south pole of the magnet slowly in and out of the loops, and observe the compass needle.
6. Move the north pole of the magnet slowly in and out of the loops, and observe the compass needle.
7. Repeat the experiment using 30 loops in the 3 m length of wire.

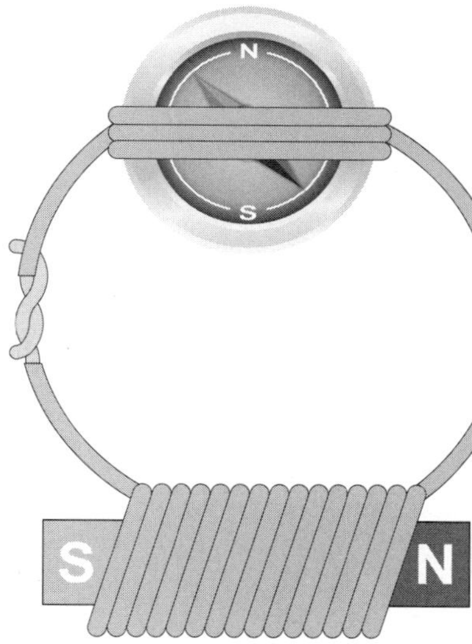

ANALYZE AND CONCLUDE:

1. How does moving the south pole of the magnet in and out of the loops of wire affect the compass needle? ______________________________

____

____

2. How does moving the north pole of the magnet in and out of the loops of wire affect the compass needle? ______________________________

____

____

Lab 7.2.2B Electromagnetic Induction

3. What effect does adding more loops of wire have on the compass needle? Why?

Name: ______________________________ Date: ______________

Lab 7.2.3A Improve the Design

PROBLEM: Using the basic components of an electric motor, how could the motion be changed or improved?

You will need:		
• 1 m length of 20-gauge insulated copper wire	• scissors • two 1.5 V batteries • masking tape	• 4 disk magnets

Steps:

1. Using the scissors, carefully remove the insulation from the ends of the wire.
2. Tape a battery or two near the edge of a table. If using two batteries, tape them in series with the positive end of one battery touching the negative end of the other battery.
3. Near each battery's terminal, tape the ends of the wire to the table. Allow the remainder of the wire to dangle over the table edge in a loop.
4. Over the side of the table, hold the grouped magnets next to the bottom of the loop of wire.
5. On the top of the table, have a team member touch one end of the wire to the positive side of the battery or batteries and simultaneously touch the other end of the wire to the negative side.
6. How could you measure or describe the response of the electric motor (wire) to the magnet? ______________________________

GENERATE IDEAS:

1. Describe how an electric motor works. ______________________________

2. How can the materials be changed to improve the motion of the motor?

Magnets	Batteries	Wire

Lab 7.2.3A Improve the Design

SELECT A SOLUTION:

3. What component will be changed? ______

4. Predict what will happen when the change is made to one component. ______

BUILD THE ITEM:

5. Draw a diagram of the changed motor and components. Be sure to label the parts.

6. Test the changed motor.

EVALUATE:

7. How did the action of the motor change? ______

Name: ______________________ Date: ____________

Lab 7.2.3A Improve the Design continued

8. How would a change to a different material affect the motor? Either continue changing the material first chosen, or select another material to change. (It is important to change only one variable at a time to ensure the change in the results is related to the new variable.)

9. Conduct the investigation and record the results. ______________________

____

____

CONCLUDE AND PRESENT RESULTS:

1. In the original investigation, what happened when the magnets were placed near the wire? ______________________

2. In the final investigation, did the change support the prediction? How? __________

____

____

3. Discuss with your team member(s) how a different electric motor might be designed. Draw a diagram and label the parts.

4. How should this new design be more successful? ______________________

____

____

5. Share your ideas with the class.

Name: ____________________ Date: __________

Lab 7.2.3B Electric Motor

QUESTION: How is the power of a motor demonstrated?

HYPOTHESIS: ____________________

____

____

EXPERIMENT:

You will need:	• 5 small disk magnets	• 2 lead wires with a clip at each end
• pre-coiled copper wire	• 2 large safety pins	
• paper or plastic cup	• 6 V battery	

Steps:

1. Turn the cup upside down. Place two of the magnets on the center of the bottom (now the top) of the cup. Affix the other three magnets inside the cup underneath the other two.
2. Tape the safety pins to the cup at opposite sides so the small holes are above the cup. These will be used as the cradle to hold the coiled wire.
3. Place the coiled wire in the cradle. The coil should clear the magnets by about 2 mm. Adjust the safety pins as needed and adjust the coil so it is balanced, centered, and spins freely. If necessary, carefully reshape the coil into a circle.
4. Attach the battery to the safety pins with the two lead wires and clips.
5. Give the coil a gentle push. (If it does not work as it should, make sure the coil is balanced and the wires are connected correctly.)

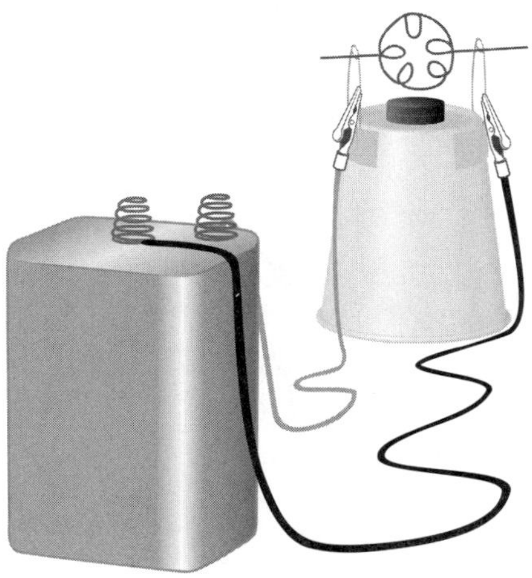

ANALYZE AND CONCLUDE:

1. What happened to the coiled wire when the lead wires were properly connected to the battery? Why? ____________________

____

____

2. How does the motor work? ____________________

____

____

____

Lab 7.2.3B Electric Motor

3. What is the benefit of an electric motor? ______________________________

____

____

____

Name: ______________________________ Date: ______________

History

WS 7.2.1A

When studying history, how is it possible to discern facts from fiction, or from personal interpretation? Much of what is known about ancient history comes from artifacts and oral records. There were no cameras to take pictures or news reporting syndicates and laws to ensure the accuracy of reported events. How is it possible to know exactly what happened or at least get some of the facts? When investigators talk to witnesses of a crime, they give more weight to the facts that are corroborated, or repeated. If two or more witnesses say the same thing, that piece of information is considered fact. If only one person reports something, that information is usually not considered as valid.

1. Research several sources that give an account of the discovery of magnetite. Keep a record of each source and list all information that is given from two or more. Then, make a list of items that are similar but not exactly the same. After you have analyzed the reports for information that appears to be valid, write a summary of what took place. Explain why your account is accurate.

"Facts" About the Discovery of Magnetite

Source	Repeated Information	Similar Information
	Where:	
	When:	
	Who:	
	What:	

2. Write your version of the discovery. ______________________________

____

____

____

3. Explain why this is accurate. ______________________________

____

____

____

Name: ______________________ Date: ____________

Generators

WS 7.2.2A

Label the parts of the generator using the student text for guidance.

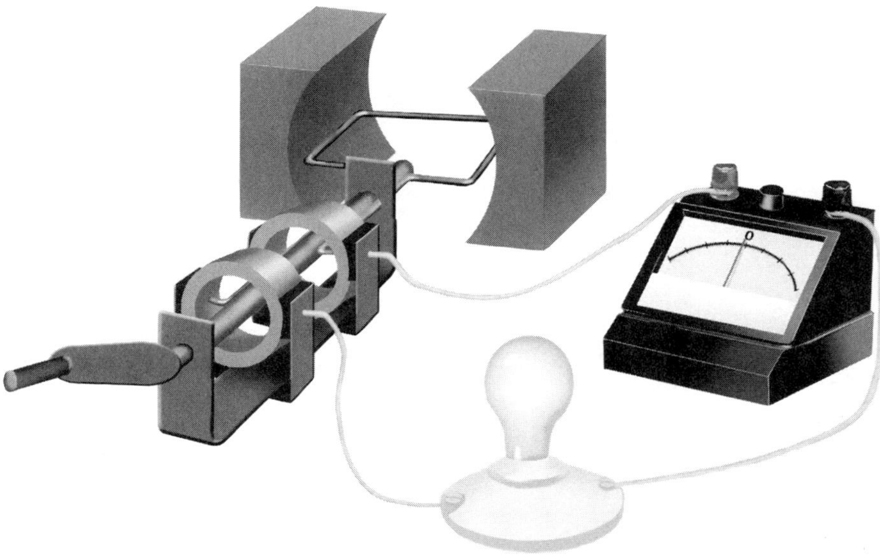

1. What is the purpose of a generator? ______________________

2. How does a generator produce alternating current? ______________________

3. What is supposed to supply the mechanical energy to the generator in the diagram? ______________________

4. What other sources are available to provide mechanical energy to power plants?

Name: ______________________________ Date: ______________

Practice

WS 7.2.3A

Use the graph to answers Questions 1–5.

Magnetic Field Around a Wire

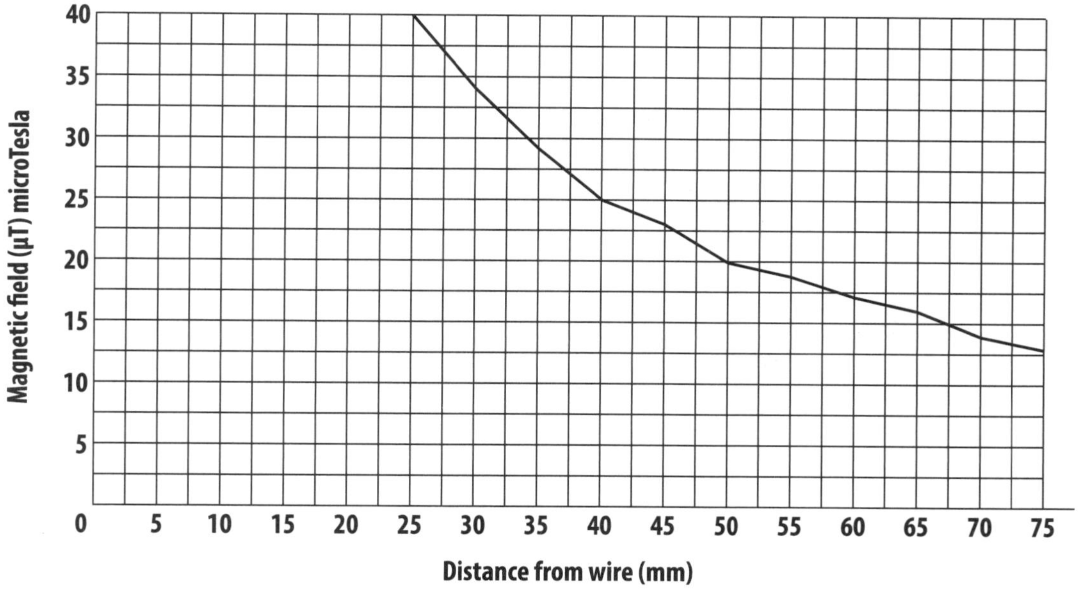

1. Estimate the magnetic field strength 45 mm from the wire. ______________

2. What distance from the wire is the magnetic field strength at 30 µT? ______________

3. By what factor does the field strength change from 60 mm to 30 mm? ______________

4. How does that compare with the change in distance? ______________

____

____

5. Does the field strength decrease more rapidly with distance closer to the wire or farther from the wire? ______________

6. A step-down transformer reduces a 2,400 V current to 120 V. If the primary coil has 500 turns of wire, how many turns of wire are in the secondary coil? ______________

7. A transformer increases the input voltage from 12 V to 12,000 V. What is the ratio of number of wire turns on the primary coil to the number of wire turns on the secondary coil? ______________

8. A step-up transformer increases the voltage from 150 V to 4,500 V. If the primary coil has 20 turns of wire, how many turns of wire are in the secondary coil?

____